Future Autonomous R

T0231189

This book provides a comprehensive overview of the past, present and future of autonomous road vehicles for professionals and students.

Split into three parts, the first section of the book brings together the key historical developments in autonomous road vehicle design and the primary explorations of the design possibilities from science fiction. This historical analysis draws upon significant test vehicles from history and explores their roles as landmarks in the evolution of the field. In addition, it also reviews the history of science fiction and outlines the key speculations about autonomous road vehicles which emerged from that world. In the second section of the book, Joseph Giacomin introduces five of the most popular future-facing speculative approaches used by designers. In doing so, he identifies the major user-facing challenges which affect ideation, product design, service design and business modelling. In the final part, science fiction prototyping is identified as the speculative approach best suited to autonomous road vehicle application. Connecting theory with practice, Giacomin provides examples of sixteen science fiction prototypes, which cover a comprehensive range of physical, psychological, sociological and ethical design challenges.

Written as an accessible guide for design practitioners and students, this book will be of use to those interested in the psychological, sociological and ethical factors involved in automotive design, human-centred design, industrial design and technology.

Joseph Giacomin is a professor of Human Centred Design at Brunel University, London, UK, where he performs research leading to products, systems and services which are physically, perceptually, cognitively and emotionally intuitive. He has worked for both the American military and the European automobile industry. He has produced more than 130 publications including the books *Automotive Human Centred Design Methods*, *Thermal – Seeing the World Through 21st Century Eyes* and, most recently,

Humans and Autonomous Vehicles. He has been a member of the editorial boards of *Ergonomics* and *International Journal of Vehicle Noise and Vibration* (IJVNV). He is a Fellow of the Chartered Institute of Ergonomics and Human Factors (CIEHF), a Fellow of the Royal Society for the encouragement of Arts, Manufactures and Commerce (RSA), a member of the Associazione Per Il Disegno Industriale (ADI) and a member of the Royal Photographic Society (RPS).

Future Autonomous Road Vehicles

JOSEPH GIACOMIN

Routledge
Taylor & Francis Group

LONDON AND NEW YORK

Designed cover image: © Max Sims

First published 2024
by Routledge
4 Park Square, Milton Park, Abingdon, Oxon OX14 4RN

and by Routledge
605 Third Avenue, New York, NY 10158

Routledge is an imprint of the Taylor & Francis Group, an informa business

© 2024 Joseph Giacomin

The right of Joseph Giacomin to be identified as author of this work has been asserted in accordance with sections 77 and 78 of the Copyright, Designs and Patents Act 1988.

British Library Cataloguing-in-Publication Data
A catalogue record for this book is available from the British Library

ISBN: 978-1-032-72422-5 (hbk)
ISBN: 978-1-032-72421-8 (pbk)
ISBN: 978-1-032-72423-2 (ebk)

DOI: 10.4324/9781032724232

Typeset in Univers
by MPS Limited, Dehradun

To Davide, may all his robots prove friendly ...

Contents

Figures

Tables

Chapter 1

Introduction

In *I, Robot* (Asimov 1940) Dr. Calvin was bringing Herbie books to learn from when the robot says: "It's the same with these books, you know, as with the others. They just don't interest me. There's nothing to your textbooks. Your science is just a mass of collected data plastered together by make-shift theory – and all so incredibly simple, that it's hardly worth bothering about. It's your fiction that interests me. Your studies of the interaction of human motives and emotions ..."

Herbie's comments reflect a general truth. Science can uncover the structure of the universe and may possibly one day explain it, but such facts do not necessarily reveal much about human behaviour or human societies. More is needed to make sense of ourselves and of our human-created world. This book is one small attempt to make sense of one small part of that human-created world, our upcoming fleet of autonomous road vehicles.

There are many tedious activities which people are happy to delegate to automated assistants, freeing time and money for other purposes. Tiring and stressful physical activities, in particular, are not missed. And for many people driving is probably one of them. Not many people enjoy driving through Piccadilly Circus during the Friday afternoon rush hour. Or on a twisty and dangerous country road in heavy rain. Or at speed on a busy motorway in fog. All through human history there has been an ongoing search for labour-saving devices and autonomous road vehicles will simply be one further such tool.

Pods, shuttles, goods delivery robots and other road vehicles are arriving which will make driving decisions autonomously. These new road vehicles will be very different ways of moving through space and time, thus they will challenge the current understanding of what machines do for people and what they mean to people. They will disrupt the current system of "automobility" which Urry (2004) defines as the ways by which humans and automobiles combine physically, socially and culturally.

DOI: 10.4324/9781032724232-1

And the disruption will be important given the central role which the automobile in particular has played in developed economies. Urry (2004) has, for example, argued that from the early 20th century onwards the automobile has been:

- the major item of individual consumption after housing which provides status to its owner/user;
- the predominant form of quasi-private mobility that subordinates other mobilities of walking, cycling, travelling by rail and so on, and reorganises how people negotiate the opportunities for, and constraints upon, work, family life, childhood, leisure and pleasure;
- the dominant culture that sustains major discourses of what constitutes the good life, what is necessary for an appropriate citizenship of mobility, and which provides literary and artistic images and symbols;
- the single most important cause of environmental resource usage;
- part of an extraordinarily powerful complex constituted through technical and social interlinkages with other industries (for example car sales and repairs, petrol refining and distribution, road-building and maintenance, roadside service areas, hotels, motels, suburban house building, marketing and advertising);
- the quintessential manufactured object produced by the leading industrial sectors and the iconic firms within 20th-century capitalism.

Given their ubiquity the automobiles have tended to disappear from our consciousness, blending into the fabric of reality as if part of nature itself. Few people alive today have experienced a world without automobiles. And fewer still have stopped to reflect upon the meanings and costs associated with automobility. Automobiles are currently as much a part of the physical world as earth, water, air and fire were for the peoples of antiquity.

But the ancien régime is about to change. Mays et al. (2022) have, for example, suggested that:

> Society is currently facing another big technological change in robotization and artificial intelligence (AI) integration, which has the same dramatic potential for change in people's lives as has the computer/internet revolution. Depending on whom you ask, robots and AI harken a third or fourth industrial revolution …

And as Oravec (2022) has noted, "Many cheerful descriptions of how robots are going to enter our workplaces and communities are also emerging in popular media, often trivialising the anxieties involved in such an injection as well as their job-related implications".

Interactions with machines, algorithms, artificially intelligent systems and robots are increasing daily. Major changes in technology and in how we live our lives are under way. And terms such as automation anxiety (Akst 2013) and robot phobia (Katz and Halpern 2014) have already entered the English language. Specific concerns which have already been voiced in relation to automation and artificial intelligence include those discussed by Jang and Nam (2022):

- user's data sharing;
- user's autonomy;
- human and AI agency;
- AI's transparency and explainability;
- AI's sociality.

Most people are becoming aware of the changes in the nature and cost of mobility. There is a shift under way from internal combustion engines to electric motors. And from human driving to autonomous driving. And from a road vehicle being a means for going somewhere to it instead being a place for doing something. The current system of automobility is giving way. Something new is arriving.

And as part of the shift there is increased leveraging of automation and of artificial intelligence which leads to technical, psychological, sociological and ethical challenges. And to legitimate causes for concern. There are, however, also legitimate reasons for optimism. And for believing that the technological, psychological, sociological, ethical and legal issues can be adequately addressed. And the author of this book is among those who believe that the challenges will be met.

In previous works the author has discussed several human-facing characteristics of autonomous road vehicles. How people might respond to an autonomous road vehicle's anthropomorphism, name, meaning, metaphor, interactions and ethics have all been discussed (see Giacomin 2023). The previous works approached the matter mainly from the point of view of the humans. How humans might think and how they might act when encountering autonomous road vehicles was the focus.

But any interaction involves at least two parties. Thus this book treats the topic which was only skirted around previously, that of the autonomous road vehicles themselves. What changes with respect to human driven road vehicles? What constraints will there be upon their design? How might they best interact with people? What services might they provide? What ethical conundrums might emerge due to their autonomy? Will they be considered a new form of life? This book fills in a few of the current gaps in relation to such human-facing matters.

Breaking from traditional automobility semantics, the term "friendly neighbourhood robot" is often used in this book to denote the future autonomous road vehicle. The use of the term should not be considered a whim or a device for adding colour to the narrative. The author has deliberately chosen the term because robots they will be, with all the imaginable complexity and sophistication. They will not be "automobiles", "cars", "vans", "lorries", "buses" or any of the other currently used metaphors and stereotypes of human-driven road vehicles. And they also won't be today's experimental autonomous road vehicles. They will instead be something different.

But what is meant by friendly neighbourhood robot? Not defining the term risks creating misunderstandings. Should a current automobile which is equipped with an advanced driver assistance system (ADAS) be considered a friendly neighbourhood robot? Or is it sufficient for a road vehicle to be able to drive itself in city traffic for limited periods of time to be considered a friendly neighbourhood robot? Or is it a friendly neighbourhood robot only once it is capable of talking to people in natural language and capable of reacting to people emotionally?

The choice of any given reference point on the spectrum between the current human-driven road vehicles and tomorrow's competent and capable companions risks appearing arbitrary. Of course. Any decision about any reference point will inevitably be a product of industrial, sociological and historical context. And any decision will be open to valid criticism from those who consider the criteria to be too generous or too severe. Nevertheless, for purposes of discussion and speculation, a datum is beneficial.

For the current purposes the point of transition from the human-driven road vehicle to the friendly neighbourhood robot is taken to be the ability to drive autonomously in normal environmental and sociological circumstances for extended periods of time. In other words, the road vehicle can be considered to be a friendly neighbourhood robot if it transports people or goods autonomously without the intervention of humans under complex road conditions and complex traffic conditions for extended periods of time.

Such a definition is of course automotive in nature; i.e. the driving component of the friendly neighbourhood robot's function is being privileged in the definition over the other functions which the robot may provide of a business, medical or entertainment nature. Revealing perhaps the professional background of the author, the definition is unashamedly "automotive". And the definition places a certain emphasis on the maturity of the product since the autonomous driving is to be maintained for extended periods of time with road conditions and with human (pedestrians, drivers, etc.) behaviours which are within the currently accepted norms.

A road vehicle which can complete a short section of motorway autonomously but which would require additional algorithms or human intervention to navigate around the scene of a major accident would not meet the definition. A road vehicle which can drive itself from Land's End to John O'Groats through typical weather conditions, typical traffic conditions and the occasional tailback or police intervention would.

The use of the term "friendly neighbourhood robot" should be considered part of a sustained narrative which addresses new issues and which addresses them in a new way. The term is likely to prove appropriate to future discourse about automation and autonomous road vehicles in ways which traditional semantics such as "automobile", "road vehicle" and others from the world of automobility will not. Eventually, the previous terms, metaphors and stereotypes will be consigned to denoting the artefacts of the past.

It is worth noting that this book discusses an artefact which does not yet fully exist. Not even in research laboratories. But which the author believes is destined to appear soon and to eventually become ubiquitous. Logic, historical trajectories and money all suggest the inevitable appearance of friendly neighbourhood robots on our streets. And when they do appear, they will only partially resemble our current road vehicle metaphors and stereotypes. They will be road vehicles, but not as we know them today.

The future friendly neighbourhood robots will share some characteristics of current human-driven road vehicles but will be something different. Something closer to the "robot" of popular imagination. Indeed, this book contains many observations from the world of robots (see Asimov and Frenkel 1985 and Ichbiah 2005 for enjoyable introductions to the world of robots) because the field of robotics has provided a wealth of information which is directly applicable to autonomous road vehicles.

Despite the many references to robots in general, however, this book is specifically about the friendly neighbourhood kind. Friendly neighbourhood robots which move people or which come to people to help them in loco. This is a clarification which the author feels is needed in order to avoid possible misunderstandings. The use of terms such as "automation", "automated system", "automated vehicle" or "autonomous vehicle" has the potential to suggest a range of self-governing land, sea and air vehicles. Without a declared focus on self-governing road vehicles, most likely carrying passengers, the author fears muddying the waters.

A further point of clarification is the distinction between "friendly neighbourhood robots" and "autonomous robots". A large number of autonomous robots come to mind which perform tasks such as battery charging, infrastructure testing, agricultural harvesting and other specialist activities in support of people. And many of them are likely to become commercial realities earlier than the friendly neighbourhood robots, due to the simplicity of their interactions with humans. But despite their early arrival and utility, this book is not about autonomous robots. The author has chosen to discuss "friendly neighbourhood robots" which transport people or which come to help people in loco. Machines which pose a very large number of human-facing design challenges.

This book focuses specifically on human-facing issues which affect the ideation, product design, service design and business modelling of the friendly neighbourhood robots. The design questions which are raised are largely psychological, sociological and ethical in nature. The point of view is unashamedly that of the human-centred designer who is asking questions about what the machine can do, should do, should cost and should mean to people.

Questions about the mechanical, electrical, data and artificial intelligence substrate of the machines are not the focus. Whether structural elements are of aluminium or steel, or whether the sensing is mostly optical or mostly thermal, or whether the power is being provided by one type of battery or another, are not subjects of this book. This book adopts the premise that the technical hurdles will be overcome, eventually. There are simply too many talented individuals already working on these matters and too much money already invested for the autonomous road vehicles to fail in these areas.

And it is worth noting that this book also does not discuss the systems of product liability and insurance which are essential if the vehicles are to be allowed on public roads. Despite the obvious need to regulate the robots and legally define their liabilities and other societal implications, such matters are dealt with comprehensively elsewhere (see, for example, Channon et al. 2019 and Turner 2018). And, once again, there are simply too many talented individuals already working on these matters and too much money already invested for the autonomous road vehicles to fail in these areas.

This book reviews the history of autonomous road vehicles and notes the key conceptual shifts which have occurred in what is meant by autonomous driving. Significant experimental vehicles from the past are described and their roles as landmarks in the evolution of the field are highlighted. This book also reviews the history of science fiction and notes the key speculations about autonomous road vehicles which have emerged from that world. A few significant autonomous road vehicles from science fiction are described and their roles as landmarks in the evolution of the imagined possibilities are highlighted. What science fiction has said, and has not said, in relation to autonomous road vehicles is noted.

This book then introduces five of the most popular future-facing speculative approaches used by designers. And it identifies one which is well suited to the friendly neighbourhood robots and to the time step into the future which is being considered. Science fiction prototypes are short stories which describe physical, psychological and social interactions with an artefact and which highlight logical and ethical conundrums. Via storytelling, they engage the imagination and stimulate "what if"? questions in the reader's mind. They are useful for exploring artefacts which can be clearly delineated and easily understood, but which are still some years in the future. In this book, science fiction prototypes are used to shed light on some of the design challenges posed by the future friendly neighbourhood robots.

Sixteen science fiction prototypes are provided in this book to explore the nature and implications of the future friendly neighbourhood robots which will start to be on our streets in ten to twenty years' time. They span a range of activities and human interactions and highlight, either explicitly or implicitly, major areas of design concern. They will hopefully add to the existing discourse and will hopefully stimulate further discussions about what the future robots should be like and about what the future robots should do. The sixteen friendly neighbourhood robots which appear in this book seem somewhat obvious opportunities to this author, and may be inevitable future developments.

The sixteen science fiction prototypes in this book have been grouped into four chapters of four robots each of which are organised along lines of service provision: transport, workplace, healthcare and entertainment. The four service areas have been repeatedly mentioned in relation to autonomous road vehicles from an early stage. And there is a degree of consensus (see, for example, SMMT 2017 and KPMG 2019) that these four areas cover a large number of the future applications of autonomous road vehicles. Given their frequency and consistency in current debate, the four service areas have been adopted as the basis for organising the science fiction prototypes in this book. The structure should prove familiar to readers and consistent with current autonomous road vehicle discourse.

The individual chapters of this book are thus the following:

Chapter 1 has introduced the aims and objectives of the book and has discussed what the book is about and what the book is not about.

The chapter structure was explained. And the term "friendly neighbour-hood robot" was defined and its implications discussed.

Chapter 2 provides an introduction to the history of autonomous road vehicles. It describes the key developments and focusses on the key conceptual shifts which have occurred in what is meant by autonomous driving. A number of significant experimental vehicles from the history of autonomous driving are described and their roles as landmarks in the evolution of the field are highlighted.

Chapter 3 introduces a set of key human-facing facts which will affect the design of the future autonomous road vehicles. And provides several human-facing definitions which are adopted throughout the remainder of book. Topics covered include autonomous road vehicle aesthetics, dynamics, behaviours, conversations, personality and trust.

Chapter 4 reviews the history of science fiction and notes several speculations about autonomous road vehicles which have emerged. Significant autonomous road vehicles from the world of science fiction are described. And what science fiction has said, and has not said, in relation to autonomous road vehicles is noted.

Chapter 5 introduces five of the best-known and most commonly encountered approaches to future-facing design speculation. Scenarios, prototypes, science fiction prototypes, design fictions and speculative design are each described. Each tends to be well suited to a different time step into the future and each was originally developed mainly for the purpose of either choosing, understanding or critiquing. The choice of science fiction prototypes as the basis for the speculations which are provided later in the book is explained.

Chapter 6 contains a set of science fiction prototypes which are about transporting people, sometimes individually and sometimes in groups. Each explores one or more characteristics and capabilities which future designers will have to consider, and which in many cases will prove difficult if not impossible to optimise. Each is a somewhat detailed speculation about the human interactions involved. And each highlights one or more logical or ethical conundrums.

Chapter 7 contains a set of science fiction prototypes which are about providing workplaces to people. Each explores one or more characteristics and capabilities which future designers will have to consider, and which in many cases will prove difficult if not impossible to optimise. Each is a somewhat detailed speculation about the human interactions involved. And each highlights one or more logical or ethical conundrums.

Chapter 8 contains a set of science fiction prototypes which are about providing healthcare to people. Each explores one or more characteristics and capabilities which future designers will have to consider, and which in many cases will prove difficult if not impossible to optimise. Each is a somewhat detailed speculation about the human interactions involved. And each highlights one or more logical or ethical conundrums.

Chapter 9 contains a set of science fiction prototypes which are about providing entertainment, sometimes individually and sometimes in groups. Each explores one or more characteristics and capabilities which future designers will have to consider, and which in many cases will prove difficult if not impossible to optimise. Each is a somewhat detailed speculation about the human interactions involved. And each highlights one or more logical or ethical conundrums.

Chapter 10 concludes the book by considering four strategic questions in relation to the future friendly neighbourhood robots: 1) With friendly neighbourhood robots what changes? 2) What challenges will the designers of the friendly neighbourhood robots face? 3) When might the friendly neighbourhood robots arrive on our streets? and 4) Will the friendly neighbourhood robots be considered a new form of life?

References

Adams, D. 2002, *The Salmon of Doubt: hitchhiking the universe one last time*, Vol. 3, Macmillan Publishers, New York, New York, USA.

Akst, D. 2013, Automation anxiety, *The Wilson Quarterly*, Vol. 37, No. 3, Summer, pp. 65–78.

Asimov, I. 1940, *I, Robot*, The Gnome Press, New York, New York, USA.

Asimov, I. and Frenkel, K.A. 1985, *Robots, Machines in Man's Image*, Harmony Books, New York, New York, USA.

Channon, M., McCormick, L. and Noussia, K. 2019, *The Law and Autonomous Vehicles*, Informa Law from Routledge, Abingdon, Oxon, UK.

Giacomin, J. 2023, *Humans and Autonomous Vehicles*, Routledge, London, UK.

Ichbiah, D. 2005, *Robots: from science fiction to technological revolution*, Harry N. Abrams Publishers, New York, USA.

Jang, S. and Nam, K.Y. 2022, Utilization of speculative design for designing human–AI interactions, *Archives of Design Research*, Vol. 35, No. 2, pp. 57–71.

Katz, J.E. and Halpern, D. 2014, Attitudes towards robots suitability for various jobs as affected robot appearance, *Behaviour & Information Technology*, Vol. 33, No. 9, pp. 941–953.

KPMG 2019, *Mobility 2030: Transforming the Mobility Landscape – how consumers and businesses can seize the benefits of the mobility revolution*, KPMG LLP, London, UK, September.

Marson, J., Ferris, K. and Dickinson, J. 2020, The Automated and Electric Vehicles Act 2018 Part 1 and Beyond: a critical review, *Statute Law Review*, Vol. 41, No. 3, pp. 395–416.

Mays, K.K., Lei, Y., Giovanetti, R. and Katz, J.E. 2022, AI as a boss? A national US survey of predispositions governing comfort with expanded AI roles in society, *AI & Society*, Vol. 37, No. 4, pp. 1587–1600.

Murphy, R.R. (ed.) 2018, *Robotics through Science Fiction: artificial intelligence explained through six classic robot short stories*, MIT Press, Cambridge, Massachusetts, USA.

Oravec, J.A. 2022, *Good Robot, Bad Robot: dark and creepy sides of robotics, autonomous vehicles, and AI*, Palgrave Macmillan, Springer Nature, Cham, Switzerland.

SMMT 2017, *Connected and Autonomous Vehicles – revolutionising mobility in society*, Society of Motor Manufacturers and Traders, London, UK.

Turner, J. 2018, *Robot Rules: regulating artificial intelligence*, Palgrave Macmillan, Cham, Switzerland.

Urry, J. 2004. The 'system' of automobility, *Theory, Culture & Society*, Vol. 21, Nos 4–5, pp. 25–39.

Chapter 2

Autonomous Road Vehicles

Past Autonomous Road Vehicles

Narratives (Maurer et al. 2016) suggest that the first attempts at removing the human driver from within a road vehicle took place in the 1920s using radio technology. Radio Corporation of America (RCA) demonstrated a radio-controlled automobile in Dayton Ohio in 1921. And in 1925 the Houdina Radio Control Company demonstrated a two-car team in New York City consisting of a controller vehicle and a controlled vehicle based on 1926 model year Chandler Motor Car Company automobiles (see Figure 2.1). The early experiments were not about autonomous driving in the sense that the automobile drove itself, but nevertheless showed that there were alternatives to the need for a human driver sat at the wheel.

As the number of road vehicles, road accidents and road deaths increased throughout the 1920s and 1930s, public attention soon began to focus on the challenges of road safety. The period saw important advances in the conceptualisation of road vehicles and of road safety. Many of the ideas appeared in a May 1938 article in *Popular Science*

Figure 2.1 The Houdina Radio Control Company two-car team.

Source: Wikimedia Commons

DOI: 10.4324/9781032724232-2

magazine (Murtfeldt 1938) titled "Highways of the Future" which described the predictions of Dr Miller McClintock, a road safety authority of the era. Among McClintock's predictions were several automotive capabilities which are familiar to people today, such as automated steering, automated braking, navigation support for the driver and centralised electronic control of traffic flow. And one of the predictions in particular, that of the use of guide wires in roadways to direct the vehicles, went on to become a focus of road safety research for several decades.

From the 1930s onwards several organisations tested roadway inductive guide-wire systems for automobiles and lorries, the goal typically being to achieve safe and secure motorway drives. Representative of the period is the General Motors (GM) Futurama exhibit which was opened to the general public at the 1939 World's Fair. The GM concept involved the use of guide wires and circuits embedded in roadways to provide electromagnetic paths for vehicles to follow. Electrically enhanced roadways which could help direct vehicles seemed to be the next big development in roadway technology.

And starting in 1953 RCA and GM jointly developed a series of prototype cars (Kilbon 1960) which used electrical circuits buried in the pavement for both directional and speed control. A cable carrying high-frequency current was used for lateral directional control (see, for example, Fenton et al. 1971 for details). And rectangular wire loops were used for providing the longitudinal speed control. As the vehicle passed over 6 by 20 foot detector loops the inductance of the loop was altered by the metal of the vehicle, thus sensing the vehicle's presence and its speed. The first complete road installation of the system was demonstrated in 1957 along a 400-foot strip of the newly constructed US Route 77 in Nebraska. The experiments continued into the early 1960s (see Figure 2.2) by which time

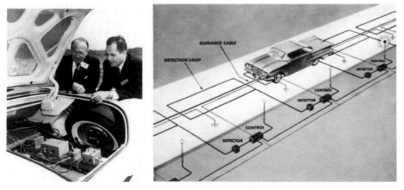

Figure 2.2 Tests of the RCA Ve-Det (Vehicle Detector) system.
Source: RCA *Electronic* Age magazine

RCA announced the commercial availability of its RCA Ve-Det (Vehicle Detector). The transistor-based system was claimed to be suitable for purposes of vehicle counting, vehicle speed measurement and traffic light operation.

The paradigm of the period is captured in images such as the advertisements published in editions of the *Saturday Evening Post* by "America's Electric Light and Power Companies" (see Figure 2.3). The ambition was to develop electronically enhanced roadways which human drivers could join, then sit back while their vehicles followed the electronic guides automatically. The paradigm was predicated on the idea that human-driving errors could be avoided if the roadway provided directional inputs to the vehicle. And the paradigm's allure revolved in large part around the promise of freeing people from the tedious and tiring driving task, permitting them to dedicate their energies to more productive and more enjoyable activities.

Experiments involving cables and other forms of electronic guidance continued well into the 1960s in several countries. For example, the UK

ELECTRICITY MAY BE THE DRIVER. One day your car may speed along an electric super-highway, its speed and steering automatically controlled by electronic devices embedded in the road. Travel will be more enjoyable. Highways will be made safe—by electricity! No traffic jams . . . no collisions . . . no driver fatigue.

Figure 2.3 Driverless Car of the Future advertisement in a 1950s edition of the *Saturday Evening Post*.

Source: The Everett Collection

Transport and Road Research Laboratory tested a modified Citroen DS (see Figure 2.4) over four miles of M4 motorway between Slough and Reading which had guidance cables installed beneath the road surface (see, for example, Dobson et al. 1974 for details).

The M4 motorway tests involved two current-carrying cables. The first was used to operate the car's steering via pick-up coils mounted on the front of the vehicle which sensed the electromagnetic field produced by the cable. Since the magnitude of the electromagnetic field depended on how far the car deviated from the cable, the field strength provided the feedback signal needed to steer the vehicle. The second cable was instead used to control the speed of the vehicle and its separation from other cars. The speed commands were issued from a central roadway controller as a voltage in the range from 0 to 10 volts, and the car used the voltage amplitude to actuate its throttle control (accelerator).

By the 1970s the costs and complexities associated with the installation of cables, sensors, transponders and other needed items of equipment for hundreds or even thousands of kilometres of roadway began to enter the public consciousness. And began to influence the self-driving research. On-board options soon began to emerge.

Advances in camera technology and in solid state electronics were improving the accuracy and lowering the cost of video-based feedback

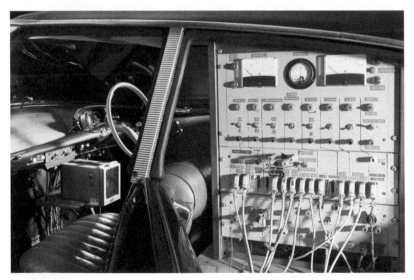

Figure 2.4 Citroen DS modified and tested by the UK Transport and Road Research Laboratory.

Source: Science Museum London

control. Several automobile manufacturers thus tested systems which used video cameras to follow roadway markings. These new camera and controller-based systems were an important change in the conceptualisation of self-driving due to shifting the driving burden from mainly the roadway to mainly the vehicle itself.

Typical of the experiments of the time are the automobiles developed by Tsukuba University and Toyota from 1977 onwards (see Figure 2.5) which used video cameras to follow lane markings. Their "Intelligent Vehicle" was the world's first automated road vehicle based on machine vision (Tsugawa 1992). The system used stereo television cameras with a field of view from 5 to 20 m ahead at a viewing angle of 40 degrees, with hard-wired logic circuits for processing the video signals to detect markings and obstacles. Intelligent Vehicle's steering was controlled using a look-up-table which had the obstacle distance and angle as inputs. Tests showed that Intelligent Vehicle could drive itself successfully at speeds of up to 30 km/h when avoiding obstacles, and at speeds of up to 50 km/h on an open road.

By this point in time the experimental vehicles had begun to drive themselves, even if only for short periods and in simple circumstances. The capabilities of their on-board electronic systems were now sufficient to drive without the aid of specialist roadway sensors or roadway guides. The experimental vehicles were detecting the features of normal roads rather than relying on signals from road infrastructure. They were starting to become autonomous.

In the 1980s in the United States the Defence Advanced Research Projects Agency (DARPA) funded projects such as Carnegie Mellon University's Navlab and ALV which led to computer-controlled vehicles involving new sensors and the now available Global Positioning

Figure 2.5 Tsukuba University automatically operated car.

Source: National Academy of Sciences

System (GPS). The widespread availability of powerful and relatively compact computers was by this point making it possible to perform highly sophisticated detection and navigation tasks on the road in real time.

Carnegie Mellon University's Navlab 1 was built in 1986 by equipping a Chevrolet van with a suite of on-board sensors, actuators and computers (see Figure 2.6). It had forward and backward facing television cameras, dedicated video hardware, a laser range finder, sonar sensors, a GPS receiver, three Sun workstations and a Warp supercomputer. By the end of the 1980s Navlab 1 was reported to be achieving speeds of up 20 mph on many real roads.

Incremental improvements were achieved with each generation of Navlab vehicle and by the time that Navlab 5 was tested in 1990 it was driving on real roads at legal speeds and requiring only occasional interventions from the humans. Navlab 5, which was based on a Pontiac Trans Sport minivan, was later used in the research team's "No Hands Across America" road trip of July 1995 where it drove the 2,850 miles from Pittsburgh to San Diego with the vehicle navigating for all but 50 of the miles.

At approximately the same time the Prometheus Project, which ran from 1987 to 1995, pooled the efforts of 14 European automotive manufacturers and of research establishments from eleven European countries. The project involved a range of mathematical models and technologies including artificial intelligence, autonomous driving,

Figure 2.6 Carnegie Mellon University's Navlab 1 vehicle.

Source: Carnegie Mellon University

vehicle-to-vehicle communications, vehicle-to-environment communications and traffic modelling. While several of the studies focussed on the paradigm of "informed driver", i.e. on assistance systems and intelligent co-pilots, several others involved fully autonomous road vehicles.

For example, in 1995 the Bundeswehr University and Mercedes-Benz developed a self-driving S-class automobile which was capable of straight-line driving and could perform lane changes. The VaMP (Versuchsfahrzeug für autonome Mobilität PKW) was equipped with inertial sensors for determining the vehicle's linear and angular accelerations and with a two-camera platform at the front and another at the rear (see Figure 2.7). VaMP carried up to 60 transputers for communications, image processing, dynamic model processing and scene analysis. Despite not using a laser scanner, radar or GPS, in one test it drove the 1,750 kilometres from Munich to Copenhagen and back at speeds of up to 177 km/h with human interventions required for only 5% of the drive.

Commercial investment in autonomous driving increased dramatically after DARPA issued its famous grand challenges. The competitions offered cash prizes and provided media visibility at a time when much progress was already being made in robotics and artificial intelligence. The first DARPA Grand Challenge competition was held in 2004 in the Mojave Desert along a 240 km route. None of the vehicles actually finished the course on the first occasion and no winner was declared. The best result had been that of the Carnegie Mellon team whose vehicle completed only 11.78 km. But the media coverage was extensive. And when the challenge was repeated in 2005 the route was this time completed by a cohort of five teams. The winning robot of 2005, Stanford University's Stanley (Thrun et al. 2006), had finished the route in 6 hours, 53 minutes and 58 seconds.

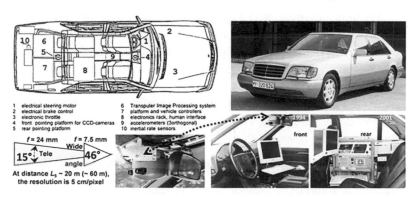

Figure 2.7 The Versuchsfahrzeug für autonome Mobilität PKW (VaMP).

Source: Wikimedia Commons

In more recent times 2017 saw Waymo's announcement that it was testing driverless automobiles without a safety driver. Then in 2018 Waymo announced that its vehicles had travelled in automated mode for over 16,000,000 km and that the total was increasing at a rate of 1,600,000 kilometres per month. By December 2018 the company was the first in the world to launch a fully autonomous commercial taxi service in Phoenix, Arizona. Autonomous vehicles were now coming close to becoming a commercial reality.

Viewed from a distance it seems that improvements in autonomous driving have been in close step with improvements in the available electronic technologies. And with the growing power and relatively low cost of current sensors and computers, it seems likely that more driving tasks and more responsibilities will be handed over to road vehicles in the coming years. There is currently no obvious sign on the horizon of a major technical impediment to further growth in abilities via hardware and via software, thus no obvious reason for doubting that fully autonomous driving will be reached. As the electronic sensing, pattern recognising and decision-making abilities improve, there will inevitably come a point when the machine will drive as safely as a human.

Present Autonomous Road Vehicles

At present the concept of autonomous road vehicle is for most people represented by a traditional human-driven road vehicle which has additional capabilities which permit it to drive itself in some situations some of the time. For many people today, what comes to mind is something along the lines of the vehicle shown in Figure 2.8. It is a stereotypical automobile which is equipped with additional electronic sensors and processing equipment to enhance the ability to make driving decisions autonomously.

There is, however, also a growing realisation that self-driving means more than additional decision-making on the part of the vehicle. If the connectivity, maps, sensors, algorithms and actuators can be made safe and reliable, then there is scope for eliminating the human driver altogether and for reconsidering major elements of the vehicle. The elimination of the driver's seat, of the manual driving controls and of the human driver leads to a very different type of road vehicle. Form and function can be revisited to seek departures from traditional automotive stereotypes.

Figure 2.9 presents an example of the current departures from traditional automotive stereotypes. The robo-taxi has some features of traditional human-driven road vehicles, but exploits the opportunities

Figure 2.8 A traditional human-driven road vehicle equipped with self-driving capabilities.

Source: Image ©unitysphere/123RF.com

Figure 2.9 A non-traditional road vehicle designed around self-driving capabilities.

Source: ©artzzz/123RF.com

offered by the elimination of the manual driving controls and by the rethinking of the seating arrangements and human ingress/egress. Numerous robo-taxis are already on streets around the world as part of mobility trials. Typically in areas of major infrastructure such as train stations, sports arenas and airports where the routes tend to be short, well defined and in some cases separated from the general traffic.

Whether enhanced versions of existing human-driven automobiles or instead non-traditional road vehicles such as robo-taxis, current autonomous road vehicles are equipped with electronic systems which can control the vehicle for periods of time without human intervention. And some autonomous road vehicles are also equipped with dedicated electronic systems for interacting with other road vehicles, with passengers and with other road users such as pedestrians. Importantly, the electronic systems are more tightly connected, coordinated and integrated than what had been past automotive practice in which individual subsystems such as the engine control, steering, braking and climate control had tended to operate somewhat independently of each other.

While the components and subsystems vary in number and complexity from manufacturer to manufacturer, a logical subdivision of the systems can be suggested based on the information which is processed and on the interactions which occur with other vehicular systems, with external machines and with humans. A list of autonomous road vehicle systems which is appropriate for use by non-specialists is presented as Table 2.1.

Achieving coordinated actions from a machine which integrates so many complex systems is not a straightforward matter. The doubts are not so much in relation to technical efficiency, since well-known principles from aerodynamics, mechanics, thermodynamics and electronics can be applied to determine matters such as the work performed by a component, the level of friction involved, the aerodynamics, the thermal losses, the needed actuation forces and other such matters. Designing the road vehicle to do things is challenging, but is well within the reach of modern engineering.

Knowing exactly what the coordinated actions should be, however, is currently more challenging. What is the best driving style? Should the vehicle appear to the outside world to be driven by a human, or not? If not human-like driving, then what inspiration from the natural world or the machine world exhibits locomotion which is simple enough and intuitive enough? What manner of driving best communicates intentions and actions? And what effects might the driving style choices have on people?

Along similar lines, what references should designers choose for the interactions with the human passengers within the vehicle, and with the pedestrians and other human road users outside the vehicle? Should

TABLE 2.1 Autonomous road vehicle systems.

Perception System	Environment perception and position localisation. The system can include one or more ultrasonic sensors, optical cameras, infrared cameras, RADAR sensors, LIDAR sensors and global positioning system (GPS) receivers. Algorithms perform data filtering, data fusion and map location detection. (See Ahangar et al. 2021 and Rosique et al. 2019 for examples.)
Planning System	Mission planning, behavioural planning and motion planning. The system can include algorithms which integrate the vehicle's global map position with the sensed local distances to obstacles and landmarks, and typically involves algorithms for determining the control actions such as accelerations, decelerations, lane keeping and lane changing. (See Rosique et al. 2019, Leon and Gavrilescu 2021 and Yoon et al. 2021 for examples.)
Control System	Trajectory prediction, trajectory tracking and vehicle hardware actuation (steering, brakes, etc.). The control algorithm determines the vehicle dynamic parameters including steering angle, longitudinal acceleration and driving style, and monitors instantaneous values to maintain them within a tolerance band. (See Leon and Gavrilescu 2021 and Yoon et al. 2021 for examples.)
Communication System	Vehicle-to-vehicle (V2V), vehicle-to-infrastructure (V2I) and vehicle-to-everything (V2X) systems. Transmission and reception technologies collect driving-relevant and passenger-relevant data for use in decision-making by the vehicle and by the passengers. (See Ahangar et al. 2021 for examples.)
Internal HMI System	Human–machine interfaces for use by passengers. The system typically includes vehicle lighting, display screens, loudspeakers and other devices for informing passengers, and buttons, optical cameras, microphones and other devices for accepting requests or commands from the passengers. (See Ataya et al. 2021 and Xing et al. 2021 for examples.)
External HMI System	Human–machine interfaces for use with other road users and pedestrians. Typically includes vehicle lighting, display screens, loudspeakers and other devices for informing other road users and/or pedestrians, and optical cameras, microphones and other devices for detecting signals, warnings and requests from other road users and/or pedestrians. (See Carmona et al. 2021 for examples.)

the communications be primarily visual? Or instead acoustic? Or a mixture of both? Should the interactions be modelled on those between humans, or not? And if not human-like in nature, then what inspiration from the natural world or the machine world is simple enough and intuitive enough to serve as the model? And what effects might the communication choices have on people?

When discussing the human interactions much of the discourse of recent years has involved the levels of automation defined by the Society of Automotive Engineers (SAE 2014). The SAE definitions (see Table 2.2) describe the roles and responsibilities of the road vehicle and of the human, in steps from little automation to full driving automation.

The SAE levels of automation provide an engineering description of what a semi-autonomous or fully autonomous road vehicle can do with respect to the nominated human driver. They describe an incrementally increasing list of driving tasks which enter the domain of the road vehicle and which exit the immediate responsibility of the human driver. They are a roadmap for how the decision-making of the automation can be increased, in steps, from that of traditional human-driven vehicles to robotic partners which deal with nearly every aspect of route planning and route following.

The intermediate SAE levels of 1, 2, 3 and 4 all imply the existence of situations in which handovers of control occur from the nominated

TABLE 2.2 SAE levels of automation.

Level 0:	The automated system issues warnings and may momentarily intervene but has no sustained control of the vehicle.
Level 1	The driver and the automated system share control of the vehicle. Examples include Cruise Control, Adaptive Cruise Control, Parking Assistance, Lane Keeping Assistance and Automatic Emergency Braking.
Level 2	The automated system takes full control of the vehicle and the driver monitors the vehicle and is prepared to intervene.
Level 3	The driver can safely turn their attention away from the driving task. The automation will handle situations that call for an immediate response, but the driver must be prepared to intervene when called upon to do so. Examples include Traffic Jam Chauffeur and Automated Lane Keeping System.
Level 4	Similar to level 3 but requiring no driver attention in limited spatial areas (geofenced) or under special circumstances.
Level 5	No human-driving intervention required in any driving content.

human driver to the vehicle, or from the vehicle to the nominated human driver. Such situations are delicate in nature and can be affected by numerous internal and external factors. Safety can be compromised by delays on the part of either the human or the automation, and misunderstandings can arise. Given the possible impacts on driving safety, much research has now been performed about the nature and characteristics of such shared responsibility scenarios.

At an early stage the transitions were noted to be cognitively complex and safety critical. And over the years it has been repeatedly noted that take-over requests by the vehicle can be problematic due to the human driver being engaged in other tasks at the time or due to the time delay associated with the shifting of the cognitive scene. And, unlike the situation with the autopilot and other flight automations of aircraft, the road vehicle environment offers much less room, literally, for error. The short physical distances involved can lead to safety critical situations for even small time delays or minor errors of interpretation.

The SAE levels are one possible way of categorising the functional characteristics of the vehicle. They are incremental in nature, though they probably do not represent equal steps in terms of complexity and cost. It is important to note, however, that the levels are mostly descriptions of the road vehicle's capabilities rather than of what the human might think or do. Defining the engineering characteristics of a control handover is not the same as defining its psychological nature, sociological impact or meaning for the people involved.

A step in the direction of a more human-centred approach may be the framework of four vehicle automation modes described by Koopman (2022). The four modes (see Table 2.3) are expressed in terms of what the human can expect. While perhaps still not saying much about who might be in the vehicle or how they might wish to interact with the vehicle, the approach does nevertheless better clarify what people can expect in terms of the driving responsibilities.

Whether describing the human interactions in terms of SAE levels or instead in terms of Koopman modes, the core consideration nevertheless remains the same. Most current autonomous road vehicle discourse focuses on driving safety. Managing to keep people safe from harm is the current goal. Avoiding "death by robot" (see Oravec 2022) is, and will remain, top priority. Current discourse usually involves what Lucy Suchman (1987) refers to as "machine operation".

But there are also other, even trickier, issues to address. For example, the issue of what Suchman (1987) refers to as "following instructions". Performing an action in an accurate and precise predetermined manner is not difficult for modern machines or recent robots.

TABLE 2.3 Four vehicle automation modes described by Koopman (2022).

Driver Assistance	A licensed human driver drives while the vehicle assists. The human performs the driving task while the vehicle provides specific functions which help to increase safety and comfort.
Supervised Automation	The vehicle controls the speed and the lane keeping while the human driver performs all other matters. The human driver monitors the road conditions and intervenes when needed to maintain safety and comfort.
Vehicle Testing	The vehicle drives fully autonomously but a trained safety driver (within the vehicle or via remote control) monitors operation and intervenes if needed to maintain safety and comfort.
Autonomous Operation	The vehicle is fully autonomous and it performs all driving activities in a safe and comfortable manner. No designated human driver is needed and vehicle primary controls such as steering, throttle and brakes are optional.

Today's actuators and controllers are impressive, transcending anything which the early pioneers of the industrial revolution could possibly have imagined. But even the amazing machines of today are still somewhat rigid and somewhat limited in their ability to accept input commands. And few of them have any real ability to understand operator intentions in a manner which would permit them to interpret the objectives and change the plans where beneficial.

This is why "following instructions" will prove challenging for the autonomous road vehicles. Most current mechanical, electrical, digital and artificial intelligence technologies still rely on simplified forms of user input for their operation. Commands are provided via a semiotically small set of instructions consisting of actions rather than intentions. With most current technologies, it is not so much a matter of the human requesting a general outcome as it is instead a matter of the human commanding a specific action.

Complicating matters further, the future autonomous road vehicles are likely to perform a large number of tasks which cannot be mechanically executed in the same manner each and every time. The people involved, the location, the time, the weather and numerous other factors will inevitably influence what will be considered appropriate or inappropriate by pedestrians, passengers and other road users. Flexibility of pick-up points, drop-off points, route planning, communications and other tasks will require a degree of judgement and autonomy which is not consistent with clockwork characters.

As Suchman (1987) has observed:

> Social studies of the production and use of instructions have concentrated on the irremediable incompleteness of instructions, and the nature of the work required in order to carry them out. The problem of the instruction-follower is viewed as one of turning essentially partial descriptions of objects and actions into concrete practical activities with predictable outcomes. A general observation from these studies is that instructions rely upon the recipient's ability to do the implicit work of anchoring descriptions to concrete objects and actions.

A useful autonomous road vehicle will thus have to be more than a Braitenberg Vehicle (Braitenberg 1986) which moves around based on fixed manipulations of sensor inputs. It will have to be capable of behaviours which are more complex than those which can be achieved through simple configurations of sensors and actuators. While a Braitenberg Vehicle may exhibit behaviours resembling fear, aggression, attraction or love from appropriate interconnections of its sensors and actuators, such behaviours are unable to adapt to changes in the motivational or social environment. While complex in nature, they are clockwork.

A useful autonomous road vehicle will instead require contextual and social grounding. It will need some mechanical interconnections of its sensors and actuators, but it will also need machine learning performed on real world datasets which are considered appropriate by designers and legislators. It needs to be taught socially and ethically acceptable behaviour. Thus choosing the real world datasets which establish the limits of its autonomy, decision-making and actions will be a major part of the design effort. A student is only as good as his teacher.

Future Autonomous Road Vehicles

At present a sedan automobile is easily recognisable from a station wagon, sport utility vehicle, sports car or van. Each vehicle type involves a typical size, mass, propulsion system and suite of on-board systems. And each involves commonalities in driving characteristics, aerodynamics, styling and several other characteristics. There are usually obvious differences in the characteristics and nature of the different automotive metaphors. And they impose important constraints on the design of human-driven road vehicles.

And deviations from the familiar forms and functions are often discouraged due to the risk of confusing people. Road safety depends, in part, on familiarity and ease of use. Keeping things the same from one generation of road vehicle to the next, and across different types of road vehicle, helps simplify matters for the drivers. There are thus many constraints placed on the design of human-driven road vehicles due to the need to achieve the greatest possible familiarity and the greatest possible safety. The need for standardisation often outweighs the benefits of innovation.

However, the introduction of road vehicles which drive themselves propels us into new territory. Familiarity and routine in relation to the driving will no longer be a matter of major concern. People pass from being drivers to being passengers. Their role in ensuring the driving safety becomes less critical. Adherence to familiar forms and functions becomes less beneficial. Designers become freer to innovate.

And where human-driven road vehicles required mainly the specification of the relatively static characteristics of form and function, road vehicles which drive themselves also need to be defined in terms of the more dynamic properties of behaviour, signalling of intention and human communication. Emotional interactions with humans may also prove beneficial, perhaps necessary. And the respecting of human social conventions will form part of the design brief.

Road vehicles which drive themselves will benefit from some of the existing automotive forms, functions and stereotypes, but will also require many new ones. The well-established traditions of the human-driven road vehicles will provide information of benefit to the substrate of the future autonomous road vehicles, but will prove of less use in relation to the design of behaviours, signalling of intentions and human communications.

The main driving capabilities of the future autonomous road vehicles are likely to include:

- driving safely on highways;
- driving safely on country roads;
- driving safely in dense urban areas;
- driving safely under common atmospheric conditions (wind, fog, rain, sleet, snow, etc.);
- dealing with detours, road works and road closures;
- dealing with road accidents which involve the vehicle itself or nearby vehicles;
- dealing with medical emergencies within the vehicle and possibly also outside the vehicle.

These capabilities will in some cases be extrapolations of the character-istics of the existing team formed by the road vehicle and its human driver. In some cases the sensing and control systems can be designed by taking as the reference what the human driver currently does when guiding the vehicle. In such cases the designer's task will be to develop technological means for mimicking the capabilities of the current combined team.

But there will also be capabilities such as dealing with road variations, accidents and medical emergencies which did not previously involve the vehicle itself to a significant degree. Dealing with such situations had always been the responsibility of the human driver in the past. But dealing with such situations will be almost entirely the responsibility of the road vehicle in the future.

The future autonomous road vehicles will be responsible for many decisions which were once within the realm of the humans, with all the accompanying safety, comfort and convenience implications. Where once upon a time a road vehicle only had to be capable of actuating, in the future it will also have to be capable of deciding. Thus if the future autonomous road vehicles are to prove safe and reliable, a new body of knowledge will be required in support of design.

And looking beyond the actual driving there is also an assortment of human interactions to design for. Autonomous road vehicles cannot be released on roads until such time as their behaviours, signalling of intentions and human communications are considered safe and appro-priate. Driving behaviours, door opening behaviours and all other physical actions will have to respect human capabilities and human norms. Internal and external communications will have to be simple, efficient and respect human capabilities and norms. Emotional interac-tions with humans may prove beneficial, perhaps necessary. And human social conventions will have to be respected when initiating interactions with people, soliciting information from them, informing them, con-firming decisions to them or simply entertaining them.

The first hundred years of automotive history have clarified the physics of road vehicles and have identified many human factors issues which can affect the driving safety. And the lessons learned can be found in the design criteria and safety standards of today's automotive industry. The physical substrate of road vehicles is now relatively well defined and the many machines which circulate on our roads today are characterised by high degrees of safety, comfort and convenience.

It thus seems likely that the next hundred years of automotive history will instead clarify how a road vehicle should act, how it should interact with people, and what meanings it should have for individuals and for society. Where the physical substrate of the vehicle is relatively well understood, the mental substrate of people's interactions with the vehicle is currently still far from being fully understood. The next hundred years will involve the exploration of desirable behaviours, of where the vehicles lie on the spectrum from machine to life form, and of what they can mean to people. The coming years will be as much about properties which can't be measured scientifically as they will be about properties which can.

In a previous work Giacomin (2023) suggested a set of human-facing autonomous road vehicle properties which are not fully amenable to scientific measurement but which are important to their success. The properties of anthropomorphism, name, meaning, metaphor, interactions and degree of ethical concern all play an important role in defining the autonomous road vehicle and in determining its place in human society. The properties (see Figure 2.10) are more about people's thoughts and behaviours than they are about physical characteristics of the vehicle. Nevertheless, they strongly influence how the autonomous road vehicle relates to people and what meanings it may have for them. Inevitably, future design briefs will have to express requirements for such properties.

And it is worth noting that for each property of Figure 2.10 the designers may find themselves having to choose between leveraging existing automotive forms, functions and stereotypes, or instead opting for new fully autonomous interpretations of road transport. Some of the future autonomous road vehicles which are more incremental in nature may benefit from adhering as much as possible to the properties of human-driven road vehicles. Other, more service-orientated machines might instead benefit from more disruptive breaks with the past. In

Anthropomorphism	None Moderate Extensive
Name	Technical Hybrid Humanoid
Meaning	Functional Ritualistic Mythical
Metaphor	Automotive Hybrid Autonomous
Interactions	Automotive Hybrid Autonomous
Ethical Concerns	None Moderate Extensive

Figure 2.10 Important human-facing properties of autonomous road vehicles.

such cases the adoption of additional anthropomorphism, of a more human-like name, and more complex metaphors and interactions might prove beneficial towards emphasising the degree of autonomy and sophistication of the service provision. What to keep, and what instead to throw away, will likely be a source of design tensions for years to come.

And it is probably also worth noting that new clusters of characteristics will surely evolve as the vehicles begin to appear on streets. Logic suggests that new mental categories and new metaphors will begin to be used and will enter the English language, probably driven more by the services rendered than the actions performed. The future autonomous road vehicles will provide very different ways of moving through space and time, thus they will disrupt the current understanding of what road vehicles do for people and what they mean to people. The current system of automobility will give way to something new.

Already now while the future autonomous road vehicles are still largely in their infancy a few clusters of characteristics can be noted in the automotive press. Figure 2.11 presents four examples of metaphors which have already become common in the specialist literature. Each is determined more by people's thoughts and behaviours than by the physical properties of the vehicle. Each involves new forms, functions and stereotypes which are more perceptual, cognitive and emotional in nature. Each is driven more by the service rendered than the driving

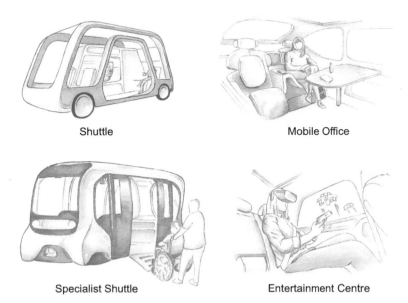

Shuttle   Mobile Office

Specialist Shuttle   Entertainment Centre

Figure 2.11 New road vehicle metaphors.

actions performed. And each is a general understanding of the vehicle which is grounded in well-known human needs and which is likely to be socially accepted.

This chapter has attempted to clarify what is meant by the term "autonomous road vehicle". The major systems which logically constitute an autonomous road vehicle were described. And the review of the history of autonomous driving highlighted key moments in the technological evolution and emphasised the major changes in paradigm. The history is one of gradual shifting of the decision-making away from the human driver. First to other humans, then to road infrastructure, then, finally, to the road vehicle itself.

This chapter has also attempted a few clarifications in relation to current autonomous road vehicle discourse. Logical points of reference including the SAE Levels and Koopman's modes were introduced, and observations were made about the implications of the shift from human-driven road vehicles to autonomous road vehicles. A major design challenge was noted, that of achieving autonomous road vehicles which are good at "following instructions".

This chapter also described a set of autonomous road vehicle properties which are not fully amenable to scientific measurement, and introduced four current autonomous road vehicle metaphors which vary across those properties. Finally, it was observed that the coming years will involve an exploration of the desirable behaviours of autonomous road vehicles, of what they can mean to people and of where they lie on the spectrum from machine to life form.

The next chapter presents instead a set of key facts which will strongly influence the design of the future autonomous road vehicles. Facts about how their appearance, movement and speech patterns will influence people's interactions with them and their trust in them. The materials discussed in the next chapter provide an introduction to a set of design issues which were only marginal concerns with human-driven road vehicles, but which are instead major concerns with the future autonomous road vehicles.

References

Ahangar, M.N., Ahmed, Q.Z., Khan, F.A. and Hafeez, M. 2021, A survey of autonomous vehicles: enabling communication technologies and challenges, *Sensors*, Vol. 21, No. 3, p. 706.

Ataya, A., Kim, W., Elsharkawy, A. and Kim, S. 2021, How to interact with a fully autonomous vehicle: naturalistic ways for drivers to intervene in the vehicle system while performing non-driving related tasks, *Sensors*, Vol. 21, No. 6, p. 2206.

Braitenberg, V. 1986, *Vehicles: experiments in synthetic psychology*, MIT Press, Cambridge, Massachusetts, USA.

Carmona, J., Guindel, C., Garcia, F. and de la Escalera, A. 2021, eHMI: review and guidelines for deployment on autonomous vehicles, *Sensors*, Vol. 21, No. 9, p. 2912.

Dobson, J.S., Penoyre, S. and Stoneman, B.G. 1974, Automatically controlled road vehicles, Conference on Control Aspects of New Forms of Guided Land Transport, Institution of Electrical Engineers, London, 28–30 August, Conference Paper 117, pp. 138–145.

Fenton, R.E., Olson, K.W. and Bender, J.G. 1971, Advances toward the Automatic Highway, No. HS-011 999. Committee on Vehicle Characteristics and presented at the 50th Annual Meeting.

Giacomin, J. 2023, *Humans and Autonomous Vehicles*, Routledge, Abingdon, Oxon, UK.

Kilbon, K. 1960, Tomorrow's Thruway Is Here Today!, *RCA Electronic Age Magazine*, Vol. 19, No. 3, Autumn, pp. 26–29.

Kim, D.S., Emerson, R.W., Naghshineh, K., Pliskow, J. and Myers, K. 2012, Impact of adding artificially generated alert sound to hybrid electric vehicles on their detectability by pedestrians who are blind, *Journal of Rehabilitation Research & Development*, Vol. 49, No.3, pp. 381–394.

Koopman, P. 2022, *How Safe Is Safe Enough? Measuring and predicting autonomous vehicle safety*, Copyright Philip Koopman, Amazon, UK.

Leon, F. and Gavrilescu, M. 2021, A review of tracking and trajectory prediction methods for autonomous driving, *Mathematics*, Vol. 9, No. 6, p. 660.

Maurer, M., Gerdes, C., Lenz, B. and Winner, H. 2016, *Autonomous Driving: technical, legal and social aspects*, Springer Nature, Berlin, Germany.

Murtfeldt, E.W. 1938, Highways of the future, *Popular Science*, Vol. 132, No. 5, May, pp. 27–29, pp. 118–119.

Oravec, J.A. 2022, *Good Robot, Bad Robot: dark and creepy sides of robotics, autonomous vehicles, and AI*, Palgrave Macmillan, Springer Nature, Cham, Switzerland.

Rosique, F., Navarro, P.J., Fernández, C. and Padilla, A. 2019, A systematic review of perception system and simulators for autonomous vehicles research, *Sensors*, Vol. 19, No. 3, p. 648.

SAE, Taxonomy 2014, Definitions for Terms Related to On-Road Motor Vehicle Automated Driving Systems, J3016, SAE International Standard.

Suchman, L.A. 1987, *Plans and Situated Actions: the problem of human–machine communication*, Cambridge University Press, Cambridge, UK.

Thrun, S., Montemerlo, M., Dahlkamp, H., Stavens, D., Aron, A., Diebel, J., Fong, P., Gale, J., Halpenny, M., Hoffmann, G. and Lau, K. 2006, Stanley: the robot that won the DARPA Grand Challenge, *Journal of Field Robotics*, Vol. 23, No. 9, pp. 661–692.

Tsugawa, S. 1992, Concept of super smart vehicle systems and their relation to advanced vehicle control systems, *Transportation Research Record*, Vol. 1358, pp. 42–49.

Xing, Y., Lv, C., Cao, D. and Hang, P. 2021, Toward human–vehicle collaboration: review and perspectives on human-centered collaborative automated driving, q*Transportation Research Part C: Emerging Technologies*, Vol. 128, p. 103199.

Yoon, Y., Chae, H. and Yi, K. 2021, High-definition map based motion planning, and control for urban autonomous driving (No. 2021-01-0098), SAE Technical Paper, Society of Automotive Engineers, Warrendale, Pennsylvania, USA.

Chapter 3

Future Autonomous Road Vehicles in Fact

Aesthetics of Future Autonomous Road Vehicles

What should a future autonomous road vehicle look like? This deceptively simple question is unfortunately not straightforward to answer. And is heavily dependent on context. A vehicle intended to provide taxi services will necessitate a different external body and different internal arrangements with respect to one intended as an ambulance or a goods delivery vehicle. And there will surely be aesthetic departures with respect to human-driven road vehicles due to the greater design freedoms provided by the elimination of the human driver's position. It is difficult at the moment to anticipate how the specific requirements of the different tasks and services might influence the aesthetics. Nevertheless, a few considerations can be suggested.

A first consideration is that anthropomorphism will play a role in the selection of the form and aesthetic. The human anthropomorphising tendency has often been noted to be an influence on the design of human-driven road vehicles. Providing forms and aesthetics which resemble those of living creatures has proved more popular than not doing so. Round headlights resembling eyes, turn signals resembling eyelashes, grilles which resemble mouths, contoured wheel arches which bring to mind arms or legs, and many other interior and exterior forms all seem to suggest that humans respond positively to a degree of anthropomorphism.

Current headlights, tail-lights, grilles and bumpers reveal a degree of uniformity which suggests a design constraint which is external to the vehicles and to the manufacturers. Something which is intrinsic to human beings. Such uniformity suggests the likely influence of the human anthropomorphising tendency. The many vehicle fronts and rears which appear to be a face, of sorts, cannot be a coincidence

Either consciously or subconsciously, automotive designers have been stimulating the human anthropomorphising tendency for much of

DOI: 10.4324/9781032724232-3

the 20th and early 21st centuries. The human anthropomorphising tendency seems to have provided at least a constraint on, and possibly a driving force of, the choice of form and aesthetic. And this for human-driven road vehicles which make few decisions by themselves and which exhibit only a very limited degree of autonomy.

And while the human anthropomorphising tendency has noticeably influenced the design of human-driven road vehicles, it is likely to influence the design of the future autonomous road vehicles even more. If it was a factor affecting the design of simple machines which were mainly about movement and safety, it should prove even more influential for machines which also perform many other complex tasks.

Being complex multifunction machines the future autonomous road vehicles will have several capabilities which are engaged with via several interactions. And the multiple capabilities might be executed differently in different contexts, and perhaps even at different times. Due to the complexity it may prove difficult for people to interpret what is going on unless assisted in some way. And perhaps the most straightforward way of helping people to interpret the actions and anticipate the events may be the adoption of a degree of anthropomorphism. Indeed, most current research seems to suggest that autonomous road vehicles are judged to be more intuitive, understandable, transparent and trustworthy if their form and aesthetic stimulate the human anthropomorphising tendency.

A second consideration emerges from observations such as that of Dörrenbächer et al. (2023) who stated: "… it is quite common to design robots to appear as either things or beings. Their form depends mostly on the application domain and sometimes on the designers' attitudes". Researchers often allude to the dichotomy of "simple mechanical tool" versus "complex anthropomorphised machine" which has been a subject of much speculation in the field of autonomous road vehicles.

Unavoidably, the form and aesthetic of the future autonomous road vehicles will be major instigators of the human anthropomorphising tendency. Visual, acoustic and other perceptual cues selectively favour some neural circuits over others. And thus some cognitive interpretations over others. Motion cues, in particular, favour the indexing to known real world examples which exhibit similar size, speed or pattern of motion.

As discussed by Giacomin (2023) the degree of anthropomorphism which is perceived from an artefact greatly affects how people interact with it. It affects their interpretation of the actions of the artefact, it affects their understanding of the artefact's intended role and it shapes the trust which is placed in the artefact. An artefact's anthropomorphism affects human subcortical and cortical processing, influencing which circuits are called into play when observing it, tracking it and predicting its next possible action.

So much so, that the degree of subjectively experienced anthropomorphism is often taken to be a differentiator between robots. One example of this can be found in the taxonomy of human–robot interaction (HRI) developed by Onnasch and Roesler (2021). Based on their review of previous research and on their own investigations they developed an encompassing taxonomy of human–robot interaction which takes into account the human, the robot, the form of interaction and the context. The Onnasch and Roesler taxonomy is shown in Figure 3.1.

Manifestations of anthropomorphism have pride of place in the taxonomy via direct evaluation under the heading of "morphology" and via indirect evaluation under the headings of "task specification" and "degree of autonomy". The options available under the heading of "morphology" are particularly revealing. "Anthropomorphic", "zoomorphic" and "technical" allude to the form and aesthetic being inspired by real world examples of a human, animal or machine nature.

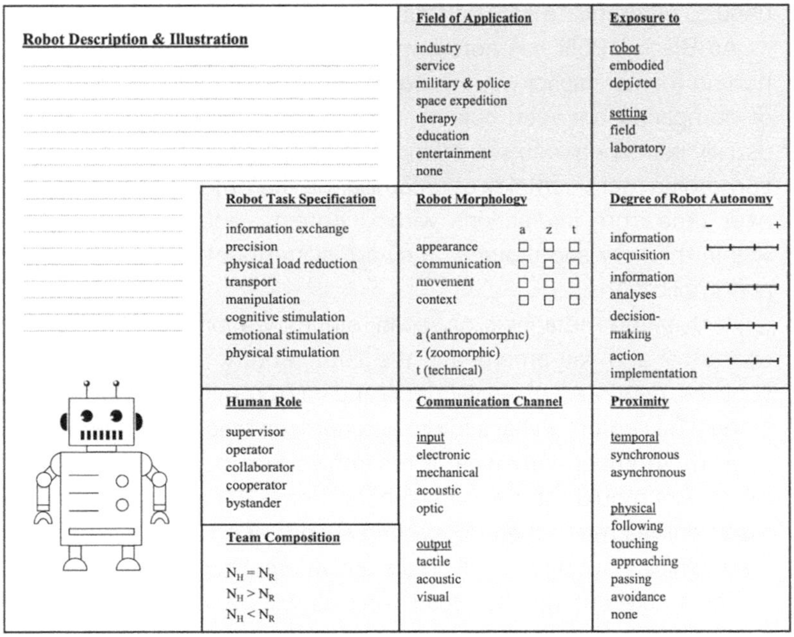

Figure 3.1 Taxonomy of human–robot interaction organised into three category clusters: interaction context (dark grey), robot (medium grey) and team classification (light grey).

Source: Onnasch and Roesler (2021)

For robots, in general, research studies have repeatedly noted (see, for example, Bernotat et al. 2017) that more contoured body shapes stimulate female connotations and associations while more angled and squared body shapes stimulate instead male connotations and associations. The perceived body shapes often also stimulate cognitive stereotypes. Leading, for example, to judging a female-shaped robot to be more appropriate for jobs which have traditionally been performed by women and a male-shaped robot more appropriate for jobs which have traditionally been performed by men. And a study by Nomura (2017) further reported that even simple gendering of a robot's name or voice is sufficient to stimulate stereotypes and to change people's reactions to the robot.

And beyond the stimulating of the human anthropomorphising tendency, the form and aesthetic of the future autonomous road vehicles will also help in the establishment of the new metaphors which will begin to be used. The form and aesthetic chosen for a future autonomous road vehicle will contribute to the activation of any relevant existing metaphors. And to the formation of any needed new ones. Form and aesthetic are currently used to differentiate between robots, thus they are likely to be used to differentiate between future autonomous vehicles.

As Black (1955) has noted, metaphor is a word used to describe a human mental capacity which serves as a basis for thinking. It is a way of comparing between objects, people or ideas, with the comparison usually being between something simple or familiar on the one hand and something more complex or unfamiliar on the other. Metaphors emerge over time from interactions within contexts, and become embedded within memory and thought. They are systems of characteristics, ideas and implications.

Just as the metaphors of sedan, station wagon, sport utility vehicle, sports car and van emerged in the 20th century for human-driven road vehicles, new ones of similar detail and interpretative power will emerge in the 21st century. What a future autonomous road vehicle looks like, and even more importantly what it resembles, will be important properties which will mould the new metaphors. Some 21st-century autonomous road vehicles may adhere to existing shapes and styles of human-driven road vehicles but others will be disruptive breaks with the past.

The future autonomous road vehicles will be very different ways of moving through space and time, thus they will disrupt the current understanding of what road vehicles look like, do for people and mean to people. And as has always been the case in design, form and aesthetic will be chosen carefully to ensure that the artefact's function and role are clearly announced.

A third consideration that is worth noting is that past automotive practice suggests that vehicle manufacturers will use form and aesthetic to emphasise the technological and futuristic nature of the future autonomous road vehicle. Such an artefact will require something new in terms of form and aesthetic. Many designers will likely opt for sleek aerodynamic shapes, high-tech materials and advanced lighting systems which can help to convey the sense of innovation and performance.

And the chosen forms and aesthetics will almost certainly also allude to any historically established styling traditions which are specific to the manufacturer. Despite the novelty, it is likely that attempts will be made, where possible, to achieve consistency and improve brand communication. Automotive styling centres are thus likely continue to play their important role in the road vehicle design process.

Dynamics of Future Autonomous Road Vehicles

The form and aesthetic of a future autonomous road vehicle will not be the only determinants of the strength of stimulation of the human anthropomorphising tendency or of the vehicle metaphor. They are major components. But not the only ones. A future autonomous road vehicle's dynamics will prove just as influential.

So what dynamics should a future autonomous road vehicle have? Also this simple question is not as straightforward to answer as it might appear. And also this one is heavily dependent on context. A vehicle intended to provide track day sports experiences necessitates a different powerplant, suspension system and aerodynamic package than one providing a shuttle service. But once again a few considerations can be suggested.

The most obvious dynamics of road vehicles are of course those associated with the driving. Straight-line acceleration, road holding in turns, overturning stability, braking distances, crash protection and other common metrics come to mind. The movement of people or goods from point A to point B is the reason why road vehicles were invented in the first place, thus it remains important going forward.

A first consideration which is worth noting in relation to the driving dynamics is that getting from point A to point B must be safe. The driving must adhere to the rules of the road, particularly those which are enshrined in law. And the driving must respect human capabilities, norms and social practices. Interactions with pedestrians and with other road users will need to be performed in manners which respect social conventions and ethical good practice. And, when something goes wrong, the vehicle must provide adequate protection in the case of a crash or other form of accident.

Given the many legislatively prescribed safety requirements for human-driven road vehicles it is unsurprising that much of the existing research in relation to autonomous road vehicles has been about driving safety (see, for example, Koopman 2022). One review by Petrović et al. (2020) reported accident statistics which suggested that current autonomous road vehicles have a reduced incidence of broadside and pedestrian collisions with respect to human-driven vehicles, but an increased incidence of rear-end collisions. The researchers suggested that human-driven vehicles were often following too closely or at unsafe speeds, most likely due to a lack of familiarity with the driving styles of the autonomous road vehicles.

And driving safety considerations lead to an interesting question which has yet to be fully explored, the possible need for a limit to the maximum speed. In many road accidents involving human-driven vehicles the driver becomes aware of the error or component failure a small number of seconds before the impact. And there is usually some attempt at steering away from the probable point of impact and some attempt at slowing the vehicle through braking. Instead, any failure of an autonomous road vehicle's planning system or control system which is large enough to lead to a crash will likely lead to one which is characterised by little or no swerving or braking before the moment of impact. Such collisions are likely to be brutal.

During any crash the momentum of the vehicle will be brought from its starting value to a value of zero. And since physical momentum is proportional to the product of the mass multiplied by the square of the velocity, the momentum which needs arresting grows with the square of the velocity. Small increases in velocity thus lead to big increases in the momentum which needs arresting, leading to big increases in the forces involved and thus big increases in the number and severity of the injuries.

Studies have investigated the passenger injuries which result from frontal impact, side impact, rollover and other accident dynamics (see, for example, Jurewicz et al. 2016). And studies have also investigated the injuries of motorcyclists (see, for example, Ding et al. 2019), bicyclists (see, for example, Kim et al. 2007), pedestrians (see, for example, Tefft 2011) and other road users. In all cases both the incidence and the severity of the injuries grew nonlinearly with the speed of the road vehicle at the moment of impact. While the number of reported injuries and their severity are not large for speeds up to about 20 km/h for pedestrian impacts and about 30km/h for vehicular impacts, the curves grow rapidly and nonlinearly from those thresholds onwards.

Given the physics, it is natural to ask if a relatively low maximum velocity (maximum speed) may need to be set for the future autonomous road vehicles. Aircraft autopilots benefit from substantial margins of error due to the ample free space around the aircraft, which can be hundreds or thousands of metres. But autonomous road vehicles have instead the potential for nearly immediate impacts due to the free space being a few centimetres or a few metres at most. The margins of error for road vehicles have always been slim, and have shrunk in recent decades as traffic has increased and the vehicles have grown in size while the roadways haven't.

Another driving dynamics safety issue is what the future autonomous road vehicles should sound like. Twentieth-century road vehicles were loud and noisy machines. Their internal combustion engines required vibration isolation, shielding and acoustically absorbent panels to reduce the sounds they emitted. Manufacturers worked hard to lower the sound pressure levels and to shape the impulses and frequency spectra to make the vehicles more comfortable. Lowering the sound levels and shaping the sound spectra was the norm. And the objective was always to make the machines more pleasing to people, and to support perceptions of quality and luxury.

But the recent introduction of electric powertrains has somewhat altered the situation. Electric vehicles are often too quiet to ensure the safety of pedestrians and of other road users. Several studies (see, for example, that of Dudenhöffer Hause 2012) have suggested that the sounds emitted by electric vehicles can be so quiet as to cause detection problems for pedestrians at crosswalks. The situation is problematic for pedestrians who have some degree of hearing impairment. And critical for pedestrians who are blind and thus may base their crossing decision mainly on the sound emitted by the oncoming road vehicle. A road vehicle which is too quiet to be heard is likely to find pedestrians dangerously crossing its path.

The problematically low levels of sound emitted by many electric road vehicles have led to calls for the use of artificially generated sounds to ensure safe detection distances (see. for example. Kim et al. 2012) for pedestrians and for other road users. And the practice of adding artificial noise has now been coded into law in several countries and by the European Union with its 2014 directive on the sound levels of motor vehicles (European Parliament and Council of the European Union 2014).

But several researchers have also noted that simply adding artificial sound is not pragmatic because it risks increasing the overall sound levels around roads. There has thus been research (see, for example, Misdariis and Cera 2013) to investigate the possibilities for achieving better acoustical detection of road vehicles though changes in the spectral composition and temporal morphology of the sounds. Rather than make the sounds louder, current efforts are directed more at making them more noticeable and more easily distinguishable.

Indeed, the term acoustic conspicuity has recently been introduced to describe the detectability of the sounds emitted by road vehicles. Research commissioned by Transport for London (TfL) has sought to develop standards of acoustic conspicuity for public vehicles such as busses (see Morgan et al. 2018). The declared objectives have been:

- create a sound that is identifiable by a wide range of the population;
- create a sound that meets the current regulations on minimum sound;
- create a sound that is not annoying or irritating to the majority of the population where the vehicles are used;
- create a sound that can be easily distinguished from other sounds, such as background sounds and other traffic.

As with TfL's work with human-driven public transport vehicles it is likely that acoustic detectability studies will have to be performed for the future autonomous road vehicles. The smaller and lighter future autonomous road vehicle will likely prove quieter and more prone to issues of acoustic conspicuity than the larger public transport vehicles. Detailed design work will be required to identify appropriate acoustic signatures. And it is likely that the acoustic conspicuity requirements will prove decisive towards determining the overall dynamics of a future autonomous road vehicle's sound emissions.

A second consideration which is worth noting in relation to the driving dynamics of the future autonomous road vehicle is the importance of the driving style. For human-driven road vehicles Elander et al. (1993) have defined the term as: "Driving style concerns the way individuals choose to drive or driving habits that have become established over a period of years. It includes choice of driving speed, threshold for overtaking, headway, and propensity to commit traffic violations. It is expected to be influenced by attitudes and beliefs relating to driving as well as more general needs and values". The concept was later extended to autonomous road vehicles by Ossig et al. (2021) who suggested: "… the definition created by Elander et al. is retained and supplemented by all further parameters of the way of driving, such as the choice of driving maneuvers. The automated driving style is defined at this point by referring this definition not to the driver but to the vehicle's automation system".

With autonomous road vehicles the umbrella term "driving style" is today used to refer to any decisions which affect the vehicle dynamics when getting from point A to point B. Once the starting point A and end point B are defined, all further decisions which affect the motion of the vehicle tend to fall under the heading of "driving style". The term thus includes all the technical criteria which are built into the vehicle's planning system and control system for use in performing actions such as stopping at a junction, moving away from a junction, negotiating a turn, accelerating, decelerating, lane keeping and the maintaining of distance from other road vehicles.

An example of research investigating autonomous road vehicle driving style is the work of Bae et al. (2020) who developed a driving controller which monitored and adjusted the vehicle's longitudinal and lateral motions. In their study the passengers preferred controller settings which maintained a low value of jerk in both the longitudinal and lateral directions, i.e. low values of the rate of change of the acceleration. Jerk, the mathematical first derivative of the acceleration, is frequently encountered in human vibration studies and is generally considered to be a measure of the abruptness of a motion. The Bae et al. results appear to suggest a passenger propensity for smoothly varying motions for autonomous road vehicles.

Another example of research investigating driving style is that of Yusof et al. (2016) whose experiments suggested that passengers preferred driving styles which were less aggressive and more defensive in nature. And from conceptually similar experiments, Basu et al. (2017) reported that passengers preferred a more defensive driving style than their own. They went on to state that, "This echoes the findings from prior work that when people are not in control of the driving they prefer lower speeds – autonomous cars are one instantiation of not being in control of the driving".

And when passenger responses were framed in terms of the trust which the autonomous road vehicle is inspiring, similar opinions emerged. For example, Ekman et al. (2019) performed a series of "Wizard of Oz" experiments whose results suggested that defensive driving styles were perceived as more trustworthy than aggressive driving styles, possibly due to the greater degree of predictability.

Behaviours of Future Autonomous Road Vehicles

The form, aesthetic and dynamics of a future autonomous road vehicle will not be the only determinants of the strength of stimulation of the human anthropomorphising tendency or of the vehicle metaphor. Its patterns of interaction with people will prove just as influential. Possibly more so.

The capable computerised companions will exhibit patterns of interaction when asking for destinations, when providing route following information, when providing digital connectivity and when performing numerous other routine duties. Whether inside with passengers or outside with pedestrians and other road users, a future autonomous road vehicle will be continually engaged in interactions which will appear to be behaviours. And in either an explicit hard-wired manner based on rules, or an implicit embedded manner based on the selection of appropriate training sets for the machine learning, or a combination of the two, the preferred patterns will need to be specified.

So what behaviours should a future autonomous road vehicle have? This question is not straightforward to answer. And it is heavily dependent on context. But once again a few considerations can be suggested.

A first consideration which is worth noting in relation to the behaviours of the future autonomous road vehicles is that selecting the best patterns of interaction for the non-driving behaviours will not prove straightforward. Scripting complete behaviours is far more complex than designing a user menu or providing a chatbot with an archive of standard responses. Choosing the right action to perform, at the right time, involving the most understandable and acceptable dynamics, is a new challenge for 21st-century designers. And it is even more difficult to script in such a way that the feedback from the humans is used to adjust, modify or eliminate the interaction. More information about human expectations and about human subjective reactions is needed for such purposes than is currently available.

One current tool which is intended as a support for such design decisions is the general catalogue of robot behaviours (Fraunhofer IAO 2022). The catalogue currently consists of forty robot interactions which are each explained individually via an example. Each is an elementary interaction between a robot and the person or persons involved. Each is described in terms of its subcomponent actions. And each has perceptual, cognitive and emotional consequences for the person or persons involved.

The robot interactions which are currently in the catalogue include: active, inactive, becoming active, becoming inactive, processing, not understanding, listening, showing, explaining, getting ready to play, encouraging good performance, questioning good performance, joyful positive feedback, displeased positive feedback, empathic negative feedback, gloating negative feedback, operation mode on, operation mode standby, operation mode off, passively available, ready, attentive, demanding action, speech recording, directing attention, input processing failed, lively, inside turn, loading animation, indication of correctness, indication of incorrectness, joyful anticipation, happy, booting, going to sleep, showing dominance, gloomy, doubtful, comforting the user and contagious energy.

Such human-distinguishable robot interactions should prove useful when organising the planning system and the control system of a future autonomous road vehicle. Regardless of the needed internal wiring and software engineering details, the human-distinguishable robot interactions provide building blocks which can be concatenated and assembled into human-distinguishable patterns. It may prove possible to use simple building blocks to construct complex behaviours.

A second consideration which is worth noting in relation to the behaviours of the future autonomous road vehicles is that choosing the right action to perform, at the right time is a major challenge, but so is choosing the manner of performing the action. The details of the physical motion, vibration or sound are important. Amplitudes and temporal dynamics can affect people as much as the action itself. Identifying the most understandable and acceptable dynamics for each action is thus another non-trivial matter. Passenger experience depends on the nature, timing and exact manner of execution of actions.

A current tool which may prove of some assistance in identifying the most understandable and acceptable dynamics for individual actions is the Robot Impression Inventory (RII) developed by Ullrich et al. (2020). It consists of nine subjective characteristics which can each take on multiple values: appearance, voice/pronunciation, life of its own/personality, movement, facial expressions, usefulness/safety, personality, fun and overall impression. The RII is a rating system which can be used to describe the execution of a given action or activity by a robot.

While not specifying the dynamics required of the physical motion, interface interaction or communicative exchange, the RII does provide a tool for subjectively rating the robot. Designers can test their ideas by prototyping their preferred dynamics and measuring the subjective impact on people. Used as a measurement tool, the RII and future similar methods may help to achieve understandable and acceptable dynamics for the individual actions of a future autonomous road vehicle.

A third consideration which is worth noting in relation to the behaviours of the future autonomous road vehicles is that beyond the right action, when to perform it and how to perform it there are also design decisions about the areas of competency to assign to the vehicle and about the degree of agency involved. What exactly will the road vehicle do autonomously? And what exactly does autonomously mean?

Designers will have to decide which perceptual, cognitive and decisional responsibilities are entirely within the realm of responsibility of the vehicle. And designers will also have to stipulate the degree of agency to assign to the vehicle when operating within its realm of responsibility. What, exactly, the term "autonomous" is referring to will have to be fully clarified. As will the options which the humans will have to override the agency, if in fact such options exist. Designers will have to define the agency in detail, ensure that all vehicle systems are compatible with that agency, and ensure that the agency is communicated to people via all actions and interactions.

For an autonomous road vehicle the term agency usually refers to its ability to sense its surroundings, analyse the data which it collects, make decisions based on that data and perform appropriate actions. An autonomous road vehicle's agency is thus about its ability to make decisions and execute actions based on its own perception of the environment.

Barandiaran et al. (2009) have defined three conditions for any object to have agency:

- Individuality: an object must be a distinguishable entity that differs from its environment.
- Interactional Asymmetry: an object must be the active source of activity in its environment.
- Normativity: an object must regulate this activity in relation to certain norms.

The second of the Barandiaran et al. conditions, interactional asymmetry, can be interpreted as suggesting that any task or activity which is routinely initiated by a future autonomous road vehicle will probably stimulate perceptions of agency in the humans. For example, if it is driving itself most of the time it will probably stimulate feelings of agency in relation to the driving. And if it should routinely initiate business, healthcare or entertainment interactions with passengers as part of its service, those interactions will likely also stimulate feelings of agency. Designers will thus find it necessary to audit the full list of activities which the vehicle will perform autonomously and to estimate the frequency of initiation of each of those activities. And designers will need to ensure that all forms of interaction with passengers, pedestrians and other road users be executed in manners which are consistent with the degree of agency. Warnings, requests and the occasional refusal should not be allowed to come as a surprise to the humans. A future autonomous road vehicle's behaviour will have to be both consistent and aligned with its intended agency.

The third of the Barandiaran et al. conditions, that of normativity, can be interpreted as suggesting the benefit of clearly defined rules of engagement. Agency in absence of rules is a recipe for misunderstandings and confusion. And while the driving rules of the road may prove relatively familiar to many passengers, pedestrians and other road users, such familiarity may not exist in relation to the other services which the future autonomous road vehicle will be providing. Designers will thus find it necessary to audit the full list of activities which the vehicle can perform autonomously and ensure that clear technical limits are established in relation to what the vehicle is allowed to do, and what it is not allowed to do. And clear rules of engagement will be required for all the human interactions including such details as how the passengers should be addressed, how to deal with conflicting requests from passengers and how to resolve inter-passenger conflicts.

A final consideration which is worth noting in relation to the behaviours of the future autonomous road vehicles is the need to take into account the psychological effects which the behaviours may have on the humans. For example, a characteristic of most machines is that they usually do nothing until commanded to action. Instead, living creatures move continuously due to physiological needs such as facilitating blood circulation and due to sensory needs such as the visual scanning of the environment. Movement, as opposed to no movement, is a major driver of the human anthropomorphising tendency. Thus the nature and frequency of a future autonomous road vehicle's movements will likely affect whether people treat it more like a machine or more like a living creature.

And future autonomous road vehicle behaviours are likely to induce psychological effects which go well beyond stimulating the human anthropomorphising tendency. For example, within the context of digital assistants for the home, Niess et al. (2018) noted that many technologies are perceived by people as either active or passive companions. Active companions initiate conversations and launch actions such as opening a calendar or reminding of an appointment. Passive companions do not normally take initiatives, waiting instead to be called for. Either way, however, feelings of companionship are stimulated which have implications on matters such as happiness, trust and dependency. Thus the initiatives taken by the future autonomous road vehicles are also likely to instigate at least some feelings of companionship in the humans. And to have at least some implications on the happiness, trust and dependency of the humans.

Conversing with Future Autonomous Road Vehicles

What should an autonomous road vehicle sound like and what should it say? Also this question is difficult to answer. Human conversations depend greatly on the characters of the people involved and on the specific context. And conversations between humans and machines are still a relatively new and rare phenomenon. Despite more than a half century of using computers and other forms of automation it is only recently that we have begun talking to them. And even more recently that we have begun expecting them to talk to us. Nevertheless, a few considerations can be suggested.

A first consideration in relation to conversing with future autonomous road vehicles is that it will be beneficial to establish the exact nature of the service provision early in the design process – and thus to establish the exact types of conversation which are needed. An overarching understanding of what the humans and the future autonomous road vehicle will be talking about will reduce the design options and simplify the design process.

And Dörrenbächer et al. (2023) have suggested three broad categories of interaction which can occur between humans and robots:

- delegating;
- cooperating;
- socialising.

For some future autonomous road vehicles such as those used by taxi services it may prove the case that the interaction between the humans and the vehicle will be mostly of a delegating type. Involving the specifying of destinations, driving styles and amenities. Such vehicles will need to communicate the available options clearly and unequivocally. And they will need to minimise the errors in the interpretation or confirmation of the passenger requests. The delegatory process will likely benefit from restrictions in language usage to narrow down options and minimise erroneous interpretations.

For other future autonomous road vehicles such as those providing medical support or entertainment the interactions may instead be more cooperative or social in nature. Such vehicles will need to probe the humans to identify the issues which need discussing or the problems which need addressing. And they will require substantial on-board information about humans and about the scientific or other knowledge of relevance to the service which is being provided. Cooperation and social interaction will likely benefit from ample linguistic, cognitive and emotional capabilities on the part of the autonomous road vehicle.

Such observations suggest the conclusion that what a future autonomous road vehicle should be saying is largely determined by the service or services which it will be providing alongside the mobility. What is appropriate at the restaurant may not be appropriate at the doctor's office. Thus service design guidelines for the general type of activity (transport, workplace, healthcare, entertainment, etc.) will prove invaluable in support of the decision-making. The world of service design will provide many useful examples of what works, and what doesn't work, for the specific type of service.

But there will still be technological issues of implementation. Technologies will have to be chosen and scripted to provide the service. Thus a second consideration in relation to conversing with future autonomous road vehicles is that there is a need to clarify as early as possible in the design process how the conversations are to be implemented. A future autonomous road vehicle will rely on a number of technologies for interacting with human passengers and with other road users. And the different technologies are based on different approaches to informing, accepting and interpreting the information.

Starting from the service which is to be provided alongside the mobility, one framework which may prove helpful towards the selection and scripting of the technologies is the system of levels defined by Nichol (2020) for artificial intelligence systems. Nichol's five levels of conversational competence for artificial intelligence systems is presented in Table 3.1.

TABLE 3.1 Five levels of conversational competence for artificial intelligence systems.

Level 1	Notification Assistants which provide simple notifications on digital devices and within individual apps.
Level 2	FAQ Assistants which allow the user to ask a simple question and obtain a simple answer. FAQ assistants are an improvement from FAQ pages with a search bar because they are often enhanced by one to two follow-up questions.
Level 3	Contextual Assistants which utilise what the user has said before by tracking where, when, how and with what urgency. Contextual tracking increases the ability to understand and to respond to unexpected inputs.
Level 4	Personalised Assistants which learn about the user over time in a manner similar to what a human might do. Remembering preferences and previous actions helps to provide personalised interfaces and to choose opportune moments to act or intervene.
Level 5	Autonomous Organisation of Assistants which knows every customer personally and which runs large parts of the company autonomously. Operations handled by the assistants might include lead generation, marketing, sales, human resources, finance and others.

Source: Adapted from Nichol (2020)

The business sector involved, the main service which is being provided and the degree of anthropomorphism selected for the vehicle will all help towards identifying the exact level of conversational competence which is required. And once chosen, that level can serve as a reference point for the design process. Maintaining consistency with respect to the chosen level of conversational competence will help towards achieving more frictionless, natural and efficient interactions.

A third consideration in relation to conversing with future autonomous road vehicles is that the vehicle's operational design domain (ODD) specification will provide some constraints on the conversations. SAE standard J3016 (2018) defines the operational design domain to be the "operating conditions under which a given driving automation system or feature thereof is specifically designed to function, including, but not limited to, environmental, geographical, and time-of-day restrictions, and/or the requisite presence or absence of certain traffic or roadway characteristics".

Current ODDs describe mostly physical conditions of the driving environment and needed vehicle driving capabilities. For example, a current ODD may contain a section which describes a road condition which the vehicle must be capable of driving in. Or it may provide guidance in relation to the sequence of actions needed to park autonomously. Current ODDs do not usually provide guidance in relation to how the vehicle should respond to user requests and do not usually provide guidance about the human interactions involved. Detailed information of a linguistic, psychological or sociological nature is currently conspicuous by its absence.

However, ODDs are meant to describe the operating environment and the vehicle capabilities at an intermediate-to-high level of abstraction. It is thus inevitable that some service design considerations will come to form part of some ODD specifications as the future autonomous road vehicles become more sophisticated and more specialised. ODDs are not likely to remain limited to only driving capabilities and driving behaviours in the future.

What should passengers be informed of? And what should be discussed? Such details will inevitably creep into some ODD specifications despite the original engineering outlook of the approach. Too little information may reduce passenger safety and passenger trust. Too much information may instead impact negatively upon the passenger experience and complicate passenger decision-making. The best human taxi drivers seem to have a knack for adjusting their conversation to the character, interests and needs of their passengers. Perhaps the future autonomous road vehicles may have to do the same.

A fourth consideration in relation to conversing with future autonomous road vehicles is that humans are likely to prefer vehicles which converse more smoothly and more intuitively. Ease of understanding and certainty of interpretation are likely to prove important. Being a "good talker" is likely to provide a significant commercial advantage.

In his book *Using Language* (1996) Herbert Clarke described how any conversation is not a sequential transfer of facts, but instead a sequence of interactions which leads to the co-creation in real time of information and meaning. He referred to language usage as being a form of joint action shaped by:

- Participants: a joint activity is carried out by two or more participants.
- Activity goals: the participants assume public roles that help determine their division of labour.
- Public goals: the participants try to establish and achieve joint public goals.
- Private goals: the participants may try individually to achieve private goals.
- Hierarchies: a joint activity ordinarily emerges as a hierarchy of joint actions.
- Procedures: the participants may exploit both conventional and non-conventional procedures.
- Boundaries: a successful joint activity has an entry and exit which are jointly engineered by the participants.
- Dynamics: joint activities may be simultaneous or intermittent, and may expand, contract, or divide in their personnel.

And Clarke (1996) emphasised that many of these factors emerge from a "common ground" which he defined as the mass of shared knowledge, beliefs and suppositions which speakers and writers use to coordinate meaning and understanding. The common ground includes knowledge of art, culture, geography, history, literature, science and many other matters which are important to humans. The common ground also includes the formal or informal linguistic rules which can be considered applicable within a given context. Choosing the best words, phrases and sentences therefore depends in part on leveraging the common ground and maintaining consistency with respect to its contents. Without reference to common ground, individual words and sentences cannot always be assigned an unambiguous meaning.

And choosing the best words, phrases and sentence structures is also presumably subject to Grice's (1975) four maxims of good practice for written or verbal exchanges between humans:

- Quantity: make your contribution as informative as is required for the current purposes of the exchange but do not make your contribution more informative than is required.
- Quality: do not say what you believe to be false and do not say that for which you lack adequate evidence.
- Relation: be relevant and stay on topic.
- Manner: avoid obscurity of expression, avoid ambiguity, be brief and be orderly.

Such is the importance of choosing the best words, phrases and sentence structures when designing any form of automation that a specific new discipline of "conversation design" has recently emerged. The discipline focusses on the rules and recommendations which automated systems should adhere to if they are to initiate, maintain and conclude good conversations with people.

In their thorough introduction to conversation design Deibel and Evanhoe (2021) highlighted the importance of following the accepted practices of the chosen language. What might seem minor matters of content or form are often instead important indicators of a person's status, authority, character or pattern of language usage. For example, Deibel and Evanhoe suggested that designers should pay attention to transition points and markers in conversations such as:

- Confirmation: ok, got it, alright, sure, etc.;
- Reaction: hmm, oops, huh, etc.;
- Creating a sequence: first, second, third, to start, then, next, finally, etc.;
- Adding information: also, what's more, in addition, etc.

Appropriate use of such markers helps to make any automated system more frictionless, informative and useful when conversing with humans. The more the machine converses like a human being, the less cognitive work is required from the people who are interacting with it. Not everyone is a roboticist or a computer programmer.

A fifth consideration in relation to conversing with future autonomous road vehicles is the likely benefit of indicators of emotion. Research supports the view that human emotions influence attention, focus, decision-making, problem-solving, goal generation and performance (Parrott 2001; Oatley et al. 2006; Evans 2019). And emotions have also been found to be contagious in the sense that people's emotions often reflect those of others with whom they are in contact. For example, Hatfield et al. (1993) described emotional contagion as "the tendency to automatically mimic and synchronize expressions, vocalizations, postures, and movements with those of another person's and, consequently, to converge emotionally".

Studies have suggested that females can be more emotionally expressive than males (Wallbott 1988; Dimberg and Lundquist 1990; Briton and Hall 1995; Bailenson et al. 2008; Kret and De Gelder 2012). And that positive emotions have been found more frequently with females (Eisenberg 1995) while negative emotions have been found more frequently with males (Evers et al. 2011; Fabes and Martin 1991). And in a road vehicle context, it has been noted that anger facilitates aggressive driving (Wells-Parker et al. 2002) and that both frustration and sadness reduce attention (Dula and Geller 2003; Lee 2010).

Given the nature of human emotions and the possibility of contagion from others it seems reasonable to expect that an automation's expressions of emotion through its voice might affect people. Indeed, emotional effects and emotional contagions have been noted in many research studies. One particularly obvious example was a study by Nass et al. (2005) who had drivers interact with a driving simulator's voice channel while driving the simulator. The results suggested that when the driver's emotions matched the vehicle's voice emotions (happy/energetic or upset/subdued) the driver spoke more to the vehicle, attended more to the road and had less than half as many accidents. It would thus seem beneficial to design autonomous road vehicle sounds and conversations with knowledge of the human emotions in mind.

Murray and Arnott (1993) have suggested that specific vocal characteristics accompany the various human emotions. Intentionally or unintentionally, human speech is modulated by the emotional state of the person who is doing the speaking. The modulation can take the form of changes in volume, pitch, speed and articulation. And any mismatches between the modulation and the context can cause confusion.

Table 3.2 lists the vocal characteristics which Murray and Arnott identified for five of the basic human emotions. The summary suggests key vocal characteristics which future autonomous road vehicles will need to modulate if emotional matching to the humans and to the context is to be achieved.

TABLE 3.2 Vocal characteristics which accompany human basic emotions.

	Fear	Disgust	Sadness	Anger	Happiness
Intensity	normal	lower	lower	higher	higher
Speech Rate	much faster	very much slower	slightly slower	slightly faster	faster or slower
Pitch Average	very much higher	very much lower	slightly lower	very much higher	much higher
Pitch Range	much wider	slightly wider	slightly narrower	much wider	much wider
Pitch Changes	normal	wide downward terminal inflections	downward inflection	abrupt on stressed syllables	smooth upward inflections
Articulation	precise	normal	slurring	tense	normal
Voice Quality	irregular voicing	grumbled chest tone	resonant	breathy chest tone	breathy blaring

Source: Adapted from Murray and Arnott (1993)

Personality of Future Autonomous Road Vehicles

And looking beyond good conversations with appropriate emotional characteristics there is also a need to establish a personality for the automation. Natural language is never simply a set of flowing words. Natural language is an interaction between living creatures who are working towards some productive purpose. Thus core components of the exchanges are the living creatures themselves. And living creatures are not living creatures without a personality.

A first consideration in relation to the personality of future autonomous road vehicles is that establishing one for the vehicle is important. The results of many research studies support this point. One obvious example is a study by Nass and Lee (2000) involving voiced book reviews and a book-buying website. Two synthesised voices (one extrovert and one introvert) and two human voices (one extrovert and one introvert) were used for the book reviews. The participants usually considered the voice to be more attractive, credible and informative when it matched their own personality (extrovert or introvert), regardless of whether it was synthetic or human. And the participants were more likely to purchase the book when the personalities matched, again regardless of whether the voice was synthetic or human.

In another study, this time involving videos of simulated autonomous road vehicles, Zhang et al. (2019) asked 443 people to interact with a vehicle which could exhibit personalities which varied across five personality traits: extroversion, agreeableness, conscientiousness, emotional stability and openness to experience. The researchers found that similarities in personality between the vehicle and the participant led to improved opinions of the vehicle's safety and other characteristics. The vehicle was more acceptable when it behaved more like the person. Over the years many similar studies have suggested that the personality which humans perceive from an automation ends up affecting their interactions with it, and their acceptance of it.

A second consideration in relation to the personality of future autonomous road vehicles is the need for criteria. Establishing a consistent personality for any form of automation requires criteria to guide the scripting and intonation of the conversations. And the target values chosen for each of the criteria should be obvious to the human interlocutors, and consistent with the nature and importance of the service which the automation is providing. One such set of criteria was suggested by Deibel and Evanhoe (2021):

- interaction goal (for example to be accurate, efficient, flexible, frictionless, guiding, personalised or trustworthy);

- level of personification (low, medium or high degree of anthropo-morphism);
- power dynamics (the authority, role and degree of intimacy);
- character traits (for example being authoritative, calm, eager, empathic, nurturing, patient, professional, sophisticated or straight-forward);
- tone (positioning the character and attitude on a spectrum such as that of formal to casual, expert to novice, warm to cool or excited to calm);
- key behaviours (the responses in situations such as talking to someone for the first time, talking to someone familiar, being interrupted, being mistaken, being asked something inappropriate, not knowing an answer, etc.).

Empirical research has confirmed the importance of several of the Deibel and Evanhoe criteria. For example, one study from the world of road vehicles (Nass and Brave 2005) varied the tone of a driving assistant technology's voice and concluded that, "Remarkably, changing the tone of a voice can strongly influence the number of accidents, the driver's perceived attention to the road, and the driver's engagement with the car". Empirical evidence points to the need for consistency in achieving and maintaining the Deibel and Evanhoe criteria if an obvious, single, personality is to be achieved for a future autonomous road vehicle.

A third consideration in relation to the personality of future autono-mous road vehicles is the benefit of acoustic matching. The character-istics of the sounds which are emitted by a speaking automation influence people's perceptions of the word meanings, information importance and personality. And several researchers have suggested the benefits which accrue from matching an automation's sound characteristics to its personality. Nass and Brave (2005) have specifically suggested four characteristics of voice sounds which help to indicate the personality of a chatbot or robot:

- volume: soft or booming;
- pitch: low-frequency deep sound or high-frequency shrill sound;
- pitch range: how much the voice alternates between low and high pitches during the conversation;
- speech rate: slow talking or fast talking.

And from their review of their findings from their empirical studies Nass and Brave (2005) concluded that: "When the voice personality and content personality were consistent, people clearly liked the voice itself more and liked the content more. That is, consistency independently improved the perception of both modalities".

A fourth consideration in relation to the personality of future autonomous road vehicles is the likely need to assign gender, and to design for it consistently. People seem to have strong preferences in relation to the gendering of automated systems and have been repeatedly shown to apply gender stereotypes and gendered expectations during conversations with them. The effects on people of machine voice gendering are so prevalent as to lead Nass and Brave (2005) to declare that "... people see so much similarity between machine gender and their own gender that females identify with 'females' and males identify with 'males'".

Consumer research has regularly noted that people overwhelmingly prefer female voices to male voices for their in-car assistants (Griggs 2011). And automotive research studies (see, for example, Large and Burnett 2013) have routinely noted that more than three-quarters of their participants preferred female voices to male voices. People are often found to consider female voices to be more polite, caring and trustworthy. As a consequence, the vast majority of today's automotive navigation systems default to a female voice when fresh from the factory (Griggs 2011).

And for autonomous road vehicles there have been several studies (see for example Nomura 2017 and Waytz et al. 2014) which have suggested changes in people's reactions, attitudes, trust and acceptance with changes in the gender of the vehicle's voice. Lee et al. (2019) performed experiments using different voice agents and found that their participants described the male voices as being more dominant and influential, and the female voices as being more trustworthy and better at discussing social matters. They also found that participants considered the autonomous road vehicle to be easier to use when there was stereotypical consistency of the voice agent's gender and its topic of conversation (i.e. male and technically focussed or female and socially focussed). And as with the in-car assistants of human-driven road vehicles, most autonomous road vehicle studies have suggested a general preference for female voices over male voices for most on-board interactions.

In good and in bad, the choice of gender and the degree of gendering of the vocal and other parameters will be major factors affecting the conversations with the future autonomous road vehicles. Over the years many studies have documented instances of humans treating inanimate objects such as televisions, computers and vehicles as if they were people (see, for example, Reeves and Nass 1996 for a review). A tendency which Turner (1987) has attributed to the fact that: "We are people. We know a lot about ourselves. And we often make sense of other things by viewing them as people too". Such anthropomorphising tendencies have been noted to be innate, instantaneous, and requiring effort and logic to counteract.

Trusting Future Autonomous Road Vehicles

Trust has always been an important issue with machines due to the safety, efficiency and economic impacts of malfunction. A machine which cannot be trusted to perform its core function in a reliable manner can quickly become a liability. Over the years features which reveal components or which communicate their momentary operating state have been routinely used in machine design to facilitate trust.

The issue of trust has been particularly felt with road vehicles due to the nature of the transport function performed. Malfunction of driving-related components can make it impossible to travel, or at least impossible to arrive on time. And complete failure of a driving-related component can have important safety implications or even be the direct cause of an accident. Access openings, fluid-level indicators and dashboard displays have all been a routine part of automotive design in the past and have all helped to stimulate trust in traditional human-driven road vehicles.

The influence of trust on the public adoption of autonomous road vehicles was recognised relatively early on. Thus marketing surveys and research studies have already been performed to identify which autonomous road vehicle characteristics might directly affect the public's trust in the vehicle. A typical example was the investigation performed by Ulahannan et al. (2017) who interacted with 36 individuals as part of an Ideas Café event at a motor museum. Thematic analysis of the information gathered from the discussions suggested the four main areas of concern presented in Table 3.3.

The concerns of Table 3.3 provide a sample of the vehicle characteristics and of the design process characteristics which are currently under discussion. Each covers an aspect of the future autonomous road vehicle, or of its design process, which can influence trust. It can be noted that people seem concerned about the nature and reliability of the new technologies, about how the various forms of human privacy will be maintained, and about the legislative and societal frameworks which will ensure justice when things go wrong.

While trust has always been a design concern the issue has assumed greater priority with the arrival of the many new forms of automation.

TABLE 3.3 Major areas of concern in relation to trust in autonomous road vehicles.

Data and Privacy	data storage
	reasons why the information is required
	acceptance data is shared
	unaware of data sharing
	differential privacy
	targeted advertising
	safety risks
Internal Interface	aesthetics
	capabilities of vehicle
	reliability
	driving-style adjustments
	customisable privacy
Society and Policy	physical privacy
	lack of driver
	coexistence of traditional and self-driving vehicles
	adoption of technology
	infrastructure
	legal and regulatory
	service and maintenance
	vehicle brand
Inclusive Design	accessibility
	age issues
	pedestrians
	involve people in design

Source: Adapted from Ulahannan (2017)

Think, for example, of the automotive navigator or of the many online services which involve interactions with chatbots. Prediction of exactly what item of information will be requested next, or of which exact warning might occur next, is not straightforward even for habitual users of the machines. With such complex multifunction machines the establishment of trust is a decisive factor in their acceptance (Yuen et al. 2020).

As machine complexity has grown and machine decisional authority has widened, the human ability to monitor machine function has been stretched and strained. And once the complexity exceeds the human ability to accurately predict the next operating state, human judgements can only be based on statistical patterns or general tendencies. There is thus a point beyond which the human judgements must pass from the realm of "prediction" to that of "trust".

Approaching the issue of trust from a psychological and managerial perspective, McKnight and Chervany (1996) have claimed that "... trust is built or destroyed through iterative reciprocal interaction". They identified six forms of trust which can occur in the case of humans but which are likely to also apply to human interactions with automation:

- trusting beliefs;
- trusting intention;
- trusting behaviour;
- system trust;
- dispositional trust;
- situational decision to trust.

The trusting of beliefs has been claimed to be the most influential of the forms because beliefs underlie intentions, which in turn underlie behaviour. Human values and human beliefs form a web of connectivities within which any given intention or any given behaviour will usually reside. Such "beliefs" establish the space within which the intentions and behaviours lie.

A first consideration in relation to trust in the future autonomous road vehicles is thus that communicating the design principles which it reflects and the operating procedures which it adheres to would seem essential if trust in beliefs, trust in intentions and system trust are to be achieved. Acceptance by the general public is likely to depend on the ability of the designers and the manufacturers to clearly articulate the vehicle's logic, principles and procedures.

A second consideration in relation to trust in the future autonomous road vehicles is that communicating relevant items of information about the instantaneous operating state of the vehicle, at least on demand if not continuously, would seem important. Providing map location, vehicle speed and any other relevant items of information about the driving and about the service provision seem important if trust in the vehicle's behaviour and intentions is to be achieved. Acceptance by the general public is likely to depend on the availability of relevant items of information, and on the clear and frictionless articulation of those items of information.

A third consideration in relation to trust in the future autonomous road vehicles is that it will be important that the vehicle assist the human decision-making. Providing the information needed to decide matters such as the destination, route, dealing with delays or dealing with emergencies is essential. Dispositional trust and situational trust will require significant amounts of accurate information about the issues at hand and about the available options. Acceptance by the general public is likely to depend on the clarity with which the humans are presented the situations and the support for human agency which the vehicle's interactions provide.

Several recent studies have delved beneath the surface of the trust construct by considering the individual psychological and sociological properties involved. Studies have attempted to identify the types of information which any form of automation will need to make available to people through the workings of its physical components, information displays, messages, scripted conversations and other human interactions. One particularly comprehensive mapping of the trust properties in relation to robots is the "trust matrix" suggested by Murphy (2019) which is shown in Figure 3.2.

The trust matrix is divided into the two major areas of "character" and "competence" which cover the manner by which the automation operates (character) and the outcomes which it achieves (competence). Both the human-facing interactions and the functional performance contribute to the trust which develops. Both will need to be highlighted to people during operation if trust is to be achieved.

The lower levels of the trust matrix list individual characteristics which usually influence judgements about "character" and "competence" to some degree. Helping people to understand the logic of an automation's "interaction scheme", its rules for ensuring "integrity", the evidence of "capability" and the evidence of "results" are all essential. Actions, interactions and items of information which highlight these characteristics should prove beneficial.

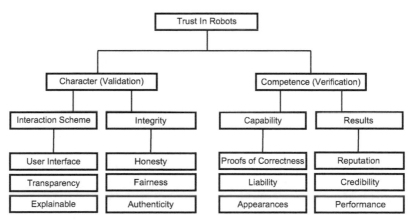

Figure 3.2 Robot trust matrix.

Source: Adapted from Murphy (2019)

As with other principles, criteria and frameworks discussed in this book the "trust matrix" operates at a somewhat high level of abstraction and provides general properties rather than detailed design criteria. Requirements such as "explainable" or "appearances" are useful references but of course do not suggest the driving style to adopt, how wide the door frame should be or the exact words which the vehicle should speak.

Nevertheless, the properties provide a reference for design discussions and for considering the possible impact of different design options. Such tools help to clarify what the future autonomous road vehicles should be doing for people and their manner of doing. And with time and practice it should prove possible to reduce the level of abstraction so as to transform the general properties into more concrete criteria of a qualitative or quantitative nature.

An early step in that direction is possibly the Trust in Self-Driving Vehicles (TSDV) subjective evaluation scale developed by Robertson (2021). The TSDV tool (see Table 3.4) is specifically intended for use with self-driving automobiles and with fully autonomous road vehicles, and operates at an intermediate level of abstraction.

The scale is based on the four factors of confidence, usability, performance and feedback which are each evaluated by means of five individual questions. A five-point Likert-type scale is used to facilitate numerical analysis such as the averaging of the individual factor scores or the reduction of the complete analysis to a single overall score (see, for example, Garcia et al. 2021).

The twenty questions of the TSDV are each expressed at an intermediate level of abstraction, thus TSDV measurements do not necessarily provide detailed design guidance. Nevertheless, designers can prototype their preferred options and deploy the tool to evaluate the impact of their design choices. Physical or virtual prototypes can be tested from a trust point of view and quantitative comparisons can be made between different designs. And, as a minimum, tools such as the TSDV shed further light on the matter of trust and provide a reference for design discussions.

A fourth consideration in relation to trust in the future autonomous road vehicles is that the twenty questions of the TSDV suggest that the driving ability is a particularly sensitive area. Several of the TSDV questions are evaluations of an aspect of the vehicle's driving. And as described in the previous chapter, the business and ethical cases for autonomous road vehicles have traditionally been based on the promise of reducing road injuries and road deaths. If future autonomous road vehicles are to be accepted they will have to prove trustworthy in the areas of driving ability and safety, first and foremost.

TABLE 3.4 Trust in Self-Driving Vehicles (TSDV) subjective evaluation scale.

	Strongly Disagree	Disagree	Neutral	Agree	Strongly Agree
Overall, I trust the self-driving car.	1	2	3	4	5
I am comfortable letting the self-driving car be in control.	1	2	3	4	5
I feel safe in the self-driving car.	1	2	3	4	5
I am not afraid of the self-driving car.	1	2	3	4	5
I am confident in the self-driving car's capabilities.	1	2	3	4	5
The self-driving car is easy to use.	1	2	3	4	5
The self-driving car is user friendly.	1	2	3	4	5
It does not require a lot of effort to use the self-driving car.	1	2	3	4	5
It is easy to get the self-driving car to do what I want.	1	2	3	4	5
I know what to do when I get in the self-driving car.	1	2	3	4	5
The self-driving car can deal with challenging road conditions.	1	2	3	4	5
The self-driving car interacts well with pedestrians.	1	2	3	4	5
The self-driving car understands traffic situations.	1	2	3	4	5
I think the self-driving car is reliable.	1	2	3	4	5

(*Continued*)

TABLE 3.4 Cont.

	Strongly Disagree	Disagree	Neutral	Agree	Strongly Agree
The self-driving car is good at driving.	1	2	3	4	5
The self-driving car communicates well.	1	2	3	4	5
The self-driving car clearly communicates what it is doing.	1	2	3	4	5
The self-driving car provides me with timely information.	1	2	3	4	5
The self-driving car 's actions are always clear to me.	1	2	3	4	5
I know what the self-driving car will do.	1	2	3	4	5

Source: Adapted from Robertson (2021)

Several studies have investigated the relationship between the driving style of an autonomous road vehicle and trust. Experiments performed by Yusof et al. (2016), Basu et al. (2017) and Ekman et al. (2019) have all confirmed that defensive driving styles are perceived as more trustworthy than aggressive driving styles, due in part to the greater degree of predictability. Reassuring pedestrians, passengers and other road users by means of a cautious and defensive driving style thus appears confirmed as a way of building trust in an autonomous road vehicle.

As noted earlier in this chapter, however, there are still design directions to be explored in terms of vehicle crashworthiness and the possible need for a maximum speed limit. The building of trust may require reassurances about how the autonomous road vehicle will protect people in the case of a failure or an accident. Beyond the behavioural manifestations during actual driving, there is a likely need for publicly communicating the safety precautions and crash protections which are provided to pedestrians, passengers and other road users.

And a final consideration in relation to trust in the future autonomous road vehicles is the benefit of anthropomorphism. For example, Waytz et al. (2014) performed a driving simulator-based study in which participants either drove a traditional automobile, rode in an autonomous road vehicle which controlled speed and steering, or rode in an autonomous road vehicle which controlled speed and steering and which was anthropomorphised by assigning it a name, gender and voice. Participant behaviour, physiological responses and self-reported psychological metrics all suggested increases in trust as the anthropomorphising features were increased.

And several studies have investigated how trust is affected by the use of different visual, acoustic and tactile approaches to interacting with the passengers. Ruijten et al. (2018), for example, investigated the benefits of conversational interfaces over traditional visual displays. Their findings suggested that conversational interfaces are anthropomorphised more, perceived as more intelligent, better liked and more trusted than traditional graphical user interfaces. Along similar lines, Forster et al. (2017) found that adding speech to an autonomous road vehicle's information system increased its degree of anthropomorphism and improved the trust in the vehicle.

And a study by Rouchitsas and Alm (2023) confirmed the benefits of anthropomorphic features when communicating with pedestrians and other road users. Pedestrians who were crossing a street were presented with an autonomous road vehicle which manifested facial emotional expressions (smile or anger) or facial conversational expressions (nod or head shake) at the front to communicate the driving intention (yield or not yield). Both forms of anthropomorphic cue were found to reduce pedestrian latency when crossing the street, with the emotional expressions proving slightly more efficient than the conversational expressions.

As a closing remark about anthropomorphism, it is worth noting the findings of Hoff and Bashir (2015) who reviewed 127 research studies of trust in automation which had been published between 2002 and 2013. Summing up the general consensus which emerged from the literature they declared that, "In order to promote greater trust and discourage automation disuse, designers should consider increasing an automated system's degree of anthropomorphism, transparency, politeness, and ease-of-use".

This chapter has discussed several facts which designers will need to keep in mind when designing future autonomous road vehicles. The effects of form and aesthetics on the human anthropomorphising tendency and on the activation of metaphors were discussed. As were the effects of the vehicle's dynamics. Design challenges in the areas of driving safety and driving style were noted, and the concept of acoustic conspicuity was introduced.

Behavioural requirements, conversational requirements, personality requirements and trust requirements were considered and several criteria and tools were introduced. The concept of agency was discussed as was the importance of expressing it clearly in the chosen actions, timings and manners of acting. The four maxims of good conversation were introduced as was the concept of conversational common ground. The importance of autonomous road vehicle personality was also highlighted as was the need to facilitate the development of trust by supporting its six individual forms.

The next chapter introduces instead the literary genre of science fiction and discusses the speculations about robots, friendly or not, which inhabit the fictional worlds of stories. It notes common themes, tropes and techniques and identifies key characteristics of the autonomous road vehicles which the stories have introduced. The societal concerns which the stories have raised, implicitly or explicitly, are also noted for the attention of designers since most of them are already present today. And the chapter draws conclusions about what has, and what has not, been speculated about autonomous road vehicles.

References

Bae, I., Moon, J., Jhung, J., Suk, H., Kim, T., Park, H., Cha, J., Kim, J., Kim, D. and Kim, S. 2020, Self-driving like a human driver instead of a Robocar: personalized comfortable driving experience for autonomous vehicles, Machine Learning for Autonomous Driving Workshop at the 33rd Conference on Neural Information Processing Systems (NeurIPS 2019), Vancouver, Canada, 14 December.

Bailenson, J.N., Pontikakis, E.D., Mauss, I.B., Gross, J.J., Jabon, M.E., Hutcherson, C.A., Nass, C. and John, O. 2008, Real-time classification of evoked emotions using facial feature tracking and physiological responses, *International Journal of Human-Computer Studies*, Vol. 66, No. 5, pp. 303–317.

Barandiaran, X.E., Di Paolo, E. and Rohde, M. 2009, Defining agency: individuality, normativity, asymmetry, and spatio-temporality in action, *Adaptive Behavior*, Vol. 17, no. 5, pp. 367–386.

Basu, C., Yang, Q., Hungerman, D., Singhal, M. and Dragan, A.D. 2017, Do you want your autonomous car to drive like you?, in Proceedings of the 2017 ACM/IEEE International Conference on Human–Robot Interaction (HRI2017), Vienna, Austria, 6–9 March, pp. 417–425.

Bernotat, J., Eyssel, F. and Sachse, J. 2017, Shape it – the influence of robot body shape on gender perception, Ninth International Conference on Social Robotics, Tsukuba, Japan, 22–24 November, pp. 75–84.

Black, M. 1955, Metaphor, Meeting of the Aristotelian Society, 21 Bedford Square, London, 23 May.

Briton, N.J. and Hall, J.A. 1995, Beliefs about female and male nonverbal communication, *Sex Roles*, Vol. 32, No. 1–2, pp. 79–90.

Clark, H.H. 1996, *Using Language*, Cambridge University Press, Cambridge, UK.

Deibel, D. and Evanhoe, R. 2021, *Conversations with Things: UX design for chat and voice*, Rosenfeld Media, New York, New York, USA.

Dimberg, U. and Lundquist, L.O. 1990, Gender differences in facial reactions to facial expressions, *Biological Psychology*, Vol. 30, No. 2, pp. 151–159.

Ding, C., Rizzi, M., Strandroth, J., Sander, U. and Lubbe, N. 2019, Motorcyclist injury risk as a function of real-life crash speed and other contributing factors, *Accident Analysis & Prevention*, Vol. 123, pp. 374–386.

Dörrenbächer, J., Ringfort-Felner, R., Neuhaus, R. and Hassenzahl, M. (eds) 2023, *Meaningful Futures with Robots: designing a new coexistence*, CRC Press, Boca Raton, Florida, USA.

Dudenhöffer, K. and Hause, L. 2012, Sound perception of electric vehicles, *ATZ Worldwide*, Vol. 114, No.3, pp. 46–50.

Dula, C.S. and Geller, E.S. 2003, Risky, aggressive, or emotional driving: addressing the need for consistent communication in research, *Journal of Safety Research*, Vol. 34, No. 5, pp. 559–566.

Eisenberg, N. and Fabes, R.A. 1995, The relation of young children's vicarious emotional responding to social competence, regulation, and emotionality, *Cognition & Emotion*, Vol. 9, No. 2–3, pp. 203–228.

Ekman, F., Johansson, M., Bligård, L.O., Karlsson, M. and Strömberg, H. 2019, Exploring automated vehicle driving styles as a source of trust information, *Transportation Research Part F: traffic psychology and behaviour*, Vol. 65, pp. 268–279.

Elander, J., West, R. and French, D. 1993, Behavioral correlates of individual differences in road-traffic crash risk: an examination of methods and findings, *Psychological Bulletin*, Vol. 113, No. 2, pp. 279–294.

European Parliament and Council of the European Union, Regulation (EU) No. 540/2014 of the European Parliament and of the Council of 16 April 2014 on the sound level of motor vehicles and of replacement silencing systems, and amending Directive 2007/46/EC and repealing Directive 70/157/EEC Text with EEA relevance.

Evans, D. 2019, *Emotion: a very short introduction* (2nd edition), Oxford University Press, Oxford, UK.

Evers, C., Fischer, A.H. and Manstead, A.S.R. 2011, Gender and emotion regulation: a social appraisal perspective on anger, in I. Nykliček, A. Vingerhoets and M. Zeelenberg (eds), *Emotion Regulation and Well-Being*, pp. 211–222, Springer, New York, New York, USA.

Fabes, R.A. and Martin, C.L. 1991, Gender and age stereotypes of emotionality, *Personality and Social Psychology Bulletin*, Vol. 17, No. 5, pp. 532–540.

Forster, Y., Naujoks, F. and Neukum, A. 2017, Increasing anthropomorphism and trust in automated driving functions by adding speech output, IEEE Intelligent Vehicles Symposium (IV), IEEE, 11–14 June, Redondo Beach, California, USA.

Fraunhofer IAO, 2022, Robot behaviour pattern Wiki, https://pattern-wiki.iao.fraunhofer.de/?page_id=427.

Garcia, K., Robertson, I. and Kortum, P. 2021, A comparison of presentation mediums for the study of trust in autonomous vehicles, *Proceedings of the Human Factors and Ergonomics Society Annual Meeting*, Vol. 65, No. 1, pp. 878–882.

Giacomin, J. 2023, *Humans and Autonomous Vehicles*, Routledge, Abingdon, Oxon, UK.

Grice, H.P. 1975, Logic and conversation, in P. Cole and J.L. Morgan (eds) *Syntax and Semantics*, Vol. 3, *Speech Acts*, New York Academic Press, New York, USA, pp. 41–58.

Griggs, B. 2011, Why computer voices are mostly female, CNN, https://edition.cnn.com/2011/10/21/tech/innovation/female-computer-voices/index.html.

Hatfield, E., Cacioppo, J.T. and Rapson, R.L. 1993, Emotional contagion, *Current Directions in Psychological Science*, Vol. 2, No. 3, pp. 96–100.

Hoff, K.A. and Bashir, M. 2015, Trust in automation: integrating empirical evidence on factors that influence trust, *Human Factors*, Vol. 57, No. 3, May, pp. 407–434.

Jurewicz, C., Sobhani, A., Woolley, J., Dutschke, J. and Corben, B. 2016, Exploration of vehicle impact speed–injury severity relationships for application in safer road design, *Transportation Research Procedia*, Vol. 14, pp. 4247–4256.

Kim, D.S., Emerson, R.W., Naghshineh, K., Pliskow, J. and Myers, K. 2012, Impact of adding artificially generated alert sound to hybrid electric vehicles on their detectability by pedestrians who are blind, *Journal of Rehabilitation Research & Development*, Vol. 49, No.3, pp. 381–394.

Kim, J.K., Kim, S., Ulfarsson, G.F. and Porrello, L.A. 2007, Bicyclist injury severities in bicycle–motor vehicle accidents, *Accident Analysis & Prevention*, Vol. 39, No.2, pp. 238–251.

Koopman, P. 2022, *How Safe Is Safe Enough? Measuring and predicting autonomous vehicle safety*, Copyright Philip Koopman, Amazon, UK.

Kret, M.E. and De Gelder, B. 2012, A review on sex differences in processing emotional signals, *Neuropsychologia*, Vol. 50, No.7, pp. 1211–1221.

Large, D.R. and Burnett, G.E. 2013, Drivers' preferences and emotional responses to satellite navigation voices, *International Journal of Vehicle Noise and Vibration*, Vol. 9, No.1/2, pp. 28–46.

Lee, S., Ratan, R. and Park, T. 2019, The voice makes the car: enhancing autonomous vehicle perceptions and adoption intention through voice agent gender and style, *Multimodal Technologies and Interaction*, Vol. 3, No.1, p. 20.

Lee, Y.C. 2010, Measuring drivers' frustration in a driving simulator, *Proceedings of the Human Factors and Ergonomics Society Annual Meeting*, September, Vol. 54, No. 19, pp. 1531–1535, Sage Publications, Thousand Oaks, California, USA.

McKnight, D.H. and Chervany, N.L. 1996, *The meanings of trust, Report MISRC 9604*, University of Minnesota MIS Research Center, Minneapolis, Minnesota, USA.

Misdariis, N. and Cera, A. 2013, Sound signature of quiet vehicles: state of the art and experience feedbacks, in INTER-NOISE Conference Proceedings, 15–18 September, Innsbruck, Austria, Institute of Noise Control Engineering, Vol. 247, No. 5, pp. 3333–3342.

Morgan, P., Edwards, A., Ainge, M., Jenkins, D., Skippon, S., Martin, P. and McCarthy, M. 2018, *The Transport for London (TfL) Bus Safety Standard: acoustic conspicuity*, TRL Limited, Wokingham, Berkshire, UK.

Murphy, R.R. 2019, *Learn AI and Human–Robot Interaction from Asimov's I, Robot Stories, Robotics through Science Fiction*, Vol. 2, Robin R. Murphy, printed by Amazon, UK.

Murray, I.R. and Arnott, J.L. 1993, Toward the simulation of emotion in synthetic speech: a review of the literature on human vocal emotion, *The Journal of the Acoustical Society of America*, Vol. 93, No. 2, pp. 1097–1108.

Nass, C.L. and Brave, S. 2005, *Wired for Speech: how voice activates and advances the human–computer relationship*, The MIT Press, Cambridge, Massachusetts, USA.

Nass, C., Jonsson, I.M., Harris, H., Reaves, B., Endo, J., Brave, S. and Takayama, L. 2005, Improving automotive safety by pairing driver emotion and car voice emotion, in *ACM Extended Abstracts Proceedings of the 2005 Conference on Human Factors in Computing Systems*, CHI 2005, Portland, Oregon, USA, 2–7 April, pp. 1973–1976.

Nass, C. and Lee, K.M. 2000, Does computer-generated speech manifest personality? An experimental test of similarity–attraction, in Proceedings of the SIGCHI Conference on Human Factors in Computing Systems, The Hague, Netherlands, 1–6 April, pp. 329–336.

Nichol, A. 2020, 5 Levels of conversational AI: 2020 update, RASA Blog, 17 June, https://rasa.com/blog/5-levels-of-conversational-ai-2020-update.

Niess, J., Diefenbach, S. and Platz, A. 2018, Moving beyond assistance: psychological qualities of digital companions, in Proceedings of the 10th Nordic Conference on Human–Computer Interaction NordiCHI'18, Oslo, Norway, 19 September–3 October, pp. 916–921.

Nomura, T. 2017, Robots and gender, *Gender and the Genome*, Vol. 1, No. 1, pp. 18–26.

Oatley, K., Keltner, D. and Jenkins, J.M. 2006, *Understanding Emotions* (2nd Edition), Blackwell Publishing, Malden, Massachusetts, USA.

Onnasch, L. and Roesler, E. 2021, A taxonomy to structure and analyze human–robot interaction, *International Journal of Social Robotics*, Vol. 13, No. 4, pp. 833–849.

Ossig, J., Cramer, S. and Bengler, K. 2021, Concept of an ontology for automated vehicle behavior in the context of human-centered research on automated driving styles, *Information*, Vol. 12, No. 1, pp. 1–14.

Parrott, W.G. (ed.) 2001, *Emotions in Social Psychology: essential readings*, Psychology Press, Taylor & Francis, Hove, East Sussex, UK.

Petrović, Ð., Mljallović, R. and Pešić, D. 2020, Traffic accidents with autonomous vehicles: type of collisions, manoeuvres and errors of conventional vehicles' drivers, *Transportation Research Procedia*, Vol. 45, pp. 161–168.

Reeves, B. and Nass, C. 1996, *The Media Equation: how people treat computers, television, and new media like real people*, Cambridge University Press, Cambridge, UK.

Ringfort-Felner, R., Laschke, M., Sadeghian, S. and Hassenzahl, M. 2022, Kiro: a design fiction to explore social conversation with voice assistants, *Proceedings of the ACM Journal on Human–Computer Interaction*, Vol. 6, No. 33, pp. 1–21.

Robertson, I. 2021, *The development and initial validation of the Trust in Self-Driving Vehicles Scale (TSDV)*, PhD thesis, Rice University, Houston, Texas, USA.

Rouchitsas, A. and Alm, H. 2023, Smiles and angry faces vs. nods and head shakes: facial expressions at the service of autonomous vehicles, *Multimodal Technologies and Interaction*, Vol. 7, No. 10, pp. 1–20.

SAE, Taxonomy 2014, Definitions for Terms Related to On-Road Motor Vehicle Automated Driving Systems, J3016, SAE International Standard.

Ruijten, P.A., Terken, J.M. and Chandramouli, S.N. 2018, Enhancing trust in autonomous vehicles through intelligent user interfaces that mimic human behavior, *Multimodal Technologies and Interaction*, Vol. 2, No. 4, pp. 1–16.

Tefft, B.C. 2011, *Impact Speed and a Pedestrian's Risk of Severe Injury or Death*, Technical Report, AAA Foundation for Traffic Safety, Washington DC, USA.

Turner, M. 1987, *Death Is the Mother of beauty: mind, metaphor, criticism*, University of Chicago Press, Chicago, Illinois, USA.

Ulahannan, A., Cain, R., Dhadyalla, G., Jennings, P., Birrell, S., Waters, M. and Mouzakitis, A. 2018, Using the ideas café to explore trust in autonomous vehicles, In A.G. Ho (ed.) *Proceedings of the AHFE 2018 International Conference on Human Factors in Communication of Design*, 21–25 July, Orlando, Florida, Springer, Cham, Switzerland, pp. 3–14.

Ullrich, D., Diefenbach, S. and Christoforakos, L. 2020, Das robot impression inventory – ein modulares instrument zur erfassung des subjektiven eindrucks von robotern, in T, Kohler, E. Schoop and N. Kahnwald (eds) *Gemeinschaften in Neuen Medien, Von Hybriden Realitäten Zu Hybriden Gemeinschaften*, Proceedings of 23rd Conference GeNeMe, TUDPress, pp. 244–249.

Wallbott, H.G. 1988, Big girls don't frown, big boys don't cry – gender differences of professional actors in communicating emotion via facial expression, *Journal of Nonverbal Behavior*, Vol. 12, No. 2, pp. 98–106.

Waytz, A., Heafner, J. and Epley, N. 2014, The mind in the machine: anthropomorphism increases trust in an autonomous vehicle, *Journal of Experimental Social Psychology*, Vol. 52, pp. 113–117.

Wells-Parker, E., Ceminsky, J., Hallberg, V., Snow, R.W., Dunaway, G., Guiling, S., Williams, M. and Anderson, B. 2002, An exploratory study of the relationship between road rage and crash experience in a representative sample of US drivers, *Accident Analysis & Prevention*, Vol. 34, No. 3, pp. 271–278.

Yuen, K.F., Wong, Y.D., Ma, F. and Wang, X. 2020, The determinants of public acceptance of autonomous vehicles: an innovation diffusion perspective, *Journal of Cleaner Production*, Vol. 270, pp. 1–13.

Yusof, N.M., Karjanto, J., Terken, J., Delbressine, F., Hassan, M.Z. and Rauterberg, M. 2016, The exploration of autonomous vehicle driving styles: preferred longitudinal, lateral, and vertical accelerations, in Proceedings of the 8th International Conference on Automotive User Interfaces and Interactive Vehicular Applications (AutomotiveUI''16), Ann Arbor, Michigan, USA, 24–26 October, pp. 245–252.

Zhang, Q., Esterwood, C., Yang, J. and Robert, L. 2019, An automated vehicle (AV) like me? The impact of personality similarities and differences between humans and AVs, AAAI Fall Symposium on Artificial Intelligence for Human–Robot Interaction, 7–9 November, Westin Arlington Gateway, Arlington, Virginia, USA.

Chapter 4

Future Autonomous Road Vehicles in Fiction

Science Fiction

While it is difficult to establish with absolute certainly who was the first person to use the term "science fiction" in a written document, there is nevertheless some consensus that it was popularised in the 1920s by the American publisher Hugo Gernsback. Dictionary entries for the term usually suggest at least three concepts:

- a form of fiction that draws imaginatively on scientific knowledge and speculation in its plot, setting and theme;
- a type of writing about imagined developments in science and their effect on life in the future;
- stories in books, magazines and films about events that take place in the future or in other parts of the universe.

Kingsley (1961) has suggested that:

> Science fiction is that class of prose narrative treating of a situation that could not arise in the world we know, but which is hypothesised on the basis of some innovation in science or technology, or pseudo-science or pseudo-technology, whether human or extraterrestrial in origin.

Seed (2011) instead suggested that:

> Science fiction has proved notoriously difficult to define. It has variously been explained as a combination of romance, science, and prophecy (Hugo Gernsback), realistic speculation about future events (Robert Heinlein), and a genre based on an imagined alternative to the reader's environment (Darko Suvin).

DOI: 10.4324/9781032724232-4

Seed went on to suggest five themes which are encountered repeatedly:

- voyages in space;
- voyages in time and alternative histories;
- encounters with aliens;
- encounters with new technologies;
- utopias and dystopias.

One attempt at narrowing the boundaries of what falls within the realm of the science fiction genre was made by Roberts (2000) who suggested that

> ... whilst Science Fiction is imaginative fiction, it does not follow that all imaginative fiction can be usefully categorised as Science Fiction. Stories in which the protagonists travel from Earth to colonies on Mars by rocket ship are usually taken to be science fiction because no such colonies, and no such available mode of transport are available to us today. But fairytales, surreal fictions (such as Andre Breton's *Nadja*, 1928), or magic realism (like Salman Rushdie's *Midnight's Children*, 1981) all involve substantive differences between the world of the text and the world the readership actually lives in, and they are not categorised as science fiction.

And further narrowing was provided by Suvin (1979) who defined science fiction as "a literary genre whose necessary and sufficient conditions are the presence and interaction of estrangement and cognition, and whose main formal device is an imaginative framework alternative to the author's empirical environment". For Suvin, science fiction stories are those which have a "novum". A technology, device or machine which is absolutely new and whose presence stimulates a different understanding of the world. The novum causes cognitive estrangement and demands that efforts be made to reach a new and stable cognitive state.

A helpful working definition of the genre was provided by the Ad Astra Institute for Science Fiction & the Speculative Imagination (2022). In their view, "Science fiction is the literature of the human species encountering change, whether it arrives via scientific discoveries, technological innovations, natural events, or societal shifts".

Over the years the genre has provided readers with:

- imaginative concepts: unconstrained by the limitations of reality, science fiction deals with imaginative and futuristic concepts that explore the boundaries of what is possible;

- technological inspiration: science fiction has been an important source of inspiration for technological and scientific advancements, with many real-world innovations being directly influenced by science fiction stories;
- world-building: science fiction often involves the creation of complete fictional worlds which have unique histories, cultures, and ecosystems, providing readers with a sense of immersion and escapism.
- social commentary: many science fiction stories use futuristic technologies and settings as a lens through which to explore contemporary social, political, and cultural issues.

And over the years the importance of the genre has been noted in many psychological, sociological and economic surveys. For example, Stockwell (2000) wrote that:

This is the most singly-identifiably popular genre of literature in the Western world. 10% of all fiction sold in Britain is science fiction … 25% of all the novels published in the US are science fiction … It is largely a paperback phenomenon, which puts it into the hands of the literate masses. On the basis of box-office receipts, more people go to watch science fiction at the cinema than any other sort of film. Science fiction scenarios create a psychological reality for the flashing lights and sound-effects of most computer games. Modern design and architecture over the past forty years have been informed by science fictional speculation about the future. Its influence can be seen in children's television programmes, games, toys and playground culture, in the corporate imagery of large companies, and in advertising and commercials for everything from cars to jeans, chocolate bars to toothpaste. Science fiction, since overtaking poetry in the 1930s, has been the most fruitful source of any area of writing in adding new words to the Oxford English Dictionary … and thus to the English language.

Science Fiction History

Identifying the roots of the science fiction genre is not straightforward. Many stories contain some elements which are considered typical of science fiction. Indeed, there are stories from as far back as antiquity which mix fact and myth to take people on a magical adventure or on a trip to a magical world. For example, *A True Story*, written by the Greek author Lucian of Samosata in the second century CE, involved space travel, alien life and interplanetary warfare.

However, despite science fiction having precursors in ancient mythology, medieval romance, post-Renaissance satire and Victorian gothic, it is usually considered a 20th-century genre. Action adventures, future visions and space operas emerged in the early 20th century as the public awareness of science and technology grew. And as improvements in paper production permitted the low cost printing of what came to be called the "pulp magazines".

It was in the early 20th century that the first major commercial manifestation of science fiction, the science fiction magazine, was born. The development is usually credited to Hugo Gernsback, an early entrepreneur of the electronics industry who later emerged as a science fiction editor and publisher. Science fiction magazines typically featured short stories, novellas and serialised novels, as well as artwork and reviews. And provided a platform for authors to showcase their ideas and stories to a dedicated readership. In these early 20th-century stories the theme of the space hero conquering the challenges posed by physics and by aliens became prevalent. As did adventures involving lasers, radioactivity, time travel and other prospected technologies.

Gernsback launched the magazine *Modern Electrics* in 1908 and soon found that its serialised stories were more popular with the readers than its technical articles. He eventually transformed *Modern Electrics* into the pulp magazine *Amazing Stories* in 1926. Highly influential, it stimulated imitators such as *Argosy*, *Astonishing Stories*, *Astounding Stories of Super-Science*, *Comet Stories*, *Cosmic Science Fiction*, *Dynamic Science Stories*, *Fantastic Adventures*, *Future Fiction*, *Galaxy SF*, *Modern Wonder*, *Planet Stories*, *Science Wonder Stories*, *Startling Stories*, *Stirring Science Stories*, *Strange Stories*, *Unknown Fantasy Fiction* and *Weird Tales*. It was through these magazines, and a few others, that many of the best known science fiction authors published their first works.

The period eventually came to be called the "Golden Age". And the exact start date came to be considered the start of John W. Campbell's influential editorship of *Astounding Science Fiction*. Campbell demanded a style of positivist storytelling which revolved around scientific achievement and human progress, with humans triumphing over all adversity. The favourable public reactions to the stories and the growing commercial success of the magazine soon attracted many of the best authors of the day, stimulated much narrative imitation and encouraged the launch of several similar magazines.

A typical Golden Age story was Isaac Asimov's 1942 *Foundation* series which chronicled the rise and fall of galactic empires and which introduced psychohistory. As was Asimov's 1950 *I, Robot* which introduced the famous "Three Laws of Robotics" and highlighted how contradictions arise from the application of fixed laws in complex settings.

Theodore Sturgeon's *More Than Human* (1953) explored the possibilities of human physical and mental evolution, touching upon several societal issues of the time including racism, classism and the fear of the unknown. And Ivan Yefremov's *Andromeda: A Space-Age Tale* (1957), which described a future interstellar communist society, highlighted the contrasts between the natural and technological worlds and explored the possible relationships between humans and artificial intelligence.

Golden Age science fiction also includes Robert A. Heinlein's *Starship Troopers* (1959) which was the first and most influential example of military science fiction. It introduced powered exoskeletons and other technologies which were later adopted by other authors. The story takes place in a future society where only those who have served in the military are considered full citizens with the right to vote and to hold public office. It explores the concepts of citizenship and democracy, and how they are not inherent rights but instead rights which must be earned. And the story also highlights the nature of war and the effects which it has on individuals and on society.

By the 1960s the science fiction stories were growing in sophistication and began exploring new themes. The period which later came to be called the "New Wave" saw writers push the boundaries of the previous publishing model which had been based on formulaic action-based stories. Reflecting the lifestyle and societal changes of the time, the new stories were incorporating avant-garde writing techniques and were exploring complex human and societal themes. With its focus on character development, psychological states and emotional themes, the New Wave science fiction broke new ground in areas of consciousness, subjectivity, hallucination and the influence of technology on people's lives.

The 1960s and 1970s saw the emergence of many young and diverse authors whose work reflected the social and cultural upheavals of the time. And from the 1970s onwards the storylines increasingly dealt with matters of identity, women's rights, politics, power and the concerns of marginalised and disempowered groups. The diversity of voices helped to broaden the appeal of science fiction and attracted many new readers to the genre.

New Wave science fiction includes Frank Herbert's 1965 *Dune* which took immersive worldbuilding to a new level of sophistication. *Dune* explored themes of ecology, religion, power and the dangers of centralised authority. The novel is often cited for its nonlinear narrative of multiple perspectives and timelines, and for its large and diverse cast of characters which each have their own motivations and agendas.

New Wave science fiction also includes Philip K. Dick's 1968 *Do Androids Dream of Electric Sheep?* which later became the inspiration for the *Blade Runner* movie franchise. The story explored the nature of empathy, identity, what it means to be a machine and what it means instead to be a human. The story's dystopian future raised questions about the ethical treatment of artificial intelligence and acted as a critique of consumerism and unchecked capitalism.

And New Wave also encompasses Ursula K. Le Guin's 1969 *The Left Hand of Darkness* which was set on a planet whose inhabitants had no fixed gender, thus challenging the traditional binary conceptions of gender of the time. The story explored several anthropological, feminist and gender themes, and critiqued power structures, particularly those related to colonialism and imperialism.

New Wave thinking stimulated the pursuit of several new themes, new directions and new subgenres. For example, William Gibson is often credited with having popularised the term "cyberspace" and his 1984 novel *Neuromancer* is often suggested to be the starting point of what has today come to be known as "cyberpunk". The novel, based in a 21st-century urban dystopia, explored virtual reality, cyberspace, artificial intelligence, altered perceptions, consciousness, self and the end of techno-capitalism.

Defined by Bruce Sterling (1986) as a "combination of lowlife and high tech", cyberpunk is usually characterised by dystopian futuristic settings and a focus on the use of technology by marginalised or criminal individuals. Common themes include virtual reality, cybernetic implants, artificial intelligence and the blurring of the boundaries between machine and human. And the stories tend to focus on characters who are outsiders, rebels or anti-heroes, struggling to survive in a world dominated by powerful corporations, corrupt governments or other oppressive forces.

Lawrence Person (1998) summarised cyberpunk with:

> Classic cyberpunk characters were marginalized, alienated loners who lived on the edge of society in generally dystopic futures where daily life was impacted by rapid technological change, an ubiquitous datasphere of computerized information, and invasive modification of the human body.

And for Goodwin (2020) cyberpunk should be considered

> ... a genre of philosophical questions about the nature of society, humankind's relationship with technology, and what it means to be

human. Growing out of a time where technology was becoming all the more omnipresent, laws governing corporations were slackening, and fashion was becoming about form over function, the questions raised by cyberpunk have only become more vital.

In line with the economic and social realities of the time, cyberpunk science fiction redrew the boundaries between the technological, the virtual and the personal. Traditional science fiction tropes such as space exploration and alien encounters were explored in new and more complex ways, leading to psychologically deeper and sociologically more sophisticated stories.

In recent times the further evolution of cyberpunk and the incorporation of new themes and tropes have produced a significant number of works which have dealt with human physical and mental enhancement, with transfer of consciousness from biological bodies to artificial devices, and with the transcending of the end of biological life. The stories have increasingly been referred to as "transhumanist" in reference to the term transhumanism which was originally used by Julian Huxley in his 1957 essay of the same name.

The philosophical and intellectual movement of transhumanism advocates the enhancement of the human condition via technologies that expand cognition and increase longevity. The characters in transhumanist science fiction stories undergo a variety of genetic engineering procedures, cybernetic enhancements, mind uploadings or other technological augmentations. In the stories the enhancements provide new abilities such as increased strength, speed or intelligence, or overcome the physical limitations of disease or aging.

But transhumanist science fiction often also includes the repercussions of the technological progress. The stories often explore philosophical and ethical issues such as the loss of individuality, the nature of the self and the problems of social inequality. Major transhumanist themes include the expanding information universe, questions about biotechnology and nanotechnology, environmental issues and the nature of post-scarcity societies.

Table 4.1 lists several well-known novels which exemplify the literary genre of science fiction. While not doing justice to the width and variety of stories which are encountered in science fiction, the list may nevertheless contain a few familiar titles which can help to illustrate the historical evolution of the genre. And of the themes and tropes which characterise it.

TABLE 4.1 Examples of literary works of science fiction.

Early Period	Lucian of Samosata's *A True Story* (2nd century CE)
	Johannes Kepler's *Somnium* (1634)
	Cyrano de Bergerac's *The Other World* (1657)
	Mary Shelley's *Frankenstein* (1818)
	Jules Verne's *Twenty Thousand Leagues Under the Sea* (1870)
	Robert Louis Stevenson's *The Strange Case of Dr. Jekyll and Mr. Hyde* (1886)
	H. G. Wells's *The Time Machine* (1895)
	H.G. Wells's *The Island of Doctor Moreau* (1896)
	H.G. Wells's *The Invisible Man* (1897)
	H.G. Wells's *The War of the Worlds* (1898)
	E. M. Forster's 'The Machine Stops' (1909)
	Edgar Rice Burroughs's *A Princess of Mars* (1912)
	Yevgeny Zamyatin's *We* (1924)
	Thea Von Harbou's *Metropolis* (1925)
	E.E. Smith's *The Skylark of Space* (1928)
	Aldous Huxley's *Brave New World* (1932)
Golden Age	Isaac Asimov's *Foundation* series (1942–1950)
	George Orwell's *1984* (1949)
	Isaac Asimov's *I, Robot* (1950)
	Ray Bradbury's *The Martian Chronicles* (1950)
	Ray Bradbury's *Fahrenheit 451* (1953)
	Theodore Sturgeon's *More Than Human* (1953)
	Arthur C. Clarke's *Childhood's End* (1953)
	Richard Matheson's *I Am Legend* (1954)
	Alfred Bester's *The Stars My Destination* (1956)
	Arthur C. Clarke's *The City and the Stars* (1956)
	Ivan Yefremov's *Andromeda: A Space-Age Tale* (1957)
	Robert A. Heinlein's *Starship Troopers* (1959)
	Kurt Vonnegut's *The Sirens of Titan* (1959)
	Walter M. Miller's *A Canticle for Leibowitz* (1959)
	Robert Heinlein's *Stranger in a Strange Land* (1961)
New Wave	Stanisław Lem's *Solaris* (1961)
	Philip K. Dick's *The Man in the High Castle* (1962)
	J.G. Ballard's *The Burning World* (1964)
	Frank Herbert's *Dune* (1965)
	Samuel R. Delany's *Babel-17* (1966)
	J.G. Ballard's *The Crystal World* (1966)
	Samuel R. Delany's *The Einstein Intersection* (1967)
	Harlan Ellison's *I Have No Mouth and I Must Scream* (1967)

(Continued)

TABLE 4.1 Cont.

	John Brunner's *Stand on Zanzibar* (1968)
	Philip K. Dick's *Do Androids Dream of Electric Sheep?* (1968)
	Vladimir Nabokov's *Ada, or Ardor* (1969)
	Philip K. Dick's *Ubik* (1969)
	Kurt Vonnegut's *Slaughterhouse-Five* (1969)
	Poul Anderson's *Tau Zero* (1970)
	J.G. Ballard's *Crash* (1973)
	Christopher Priest's *Inverted World* (1974)
	Ursula K. Le Guin's *The Dispossessed* (1974)
	John Crowley's *The Deep* (1975)
	Samuel R. Delany's *Dhalgren* (1975)
	Philip K. Dick's *A Scanner Darkly* (1977)
	Octavia E. Butler's *Kindred* (1979)
	Douglas Adams' *The Hitchhiker's Guide to the Galaxy* (1979)
Cyberpunk	William Gibson's *Burning Chrome* (1982)
	William Gibson's *Neuromancer* (1984)
	Lewis Shiner's *Frontera* (1984)
	Walter Jon Williams's *Hardwired* (1986)
	Pat Cadigan's *Mindplayers* (1987)
	Iain M. Banks's *Consider Phlebas* (1987)
	Dan Simmons's *Hyperion* (1989)
	Neal Stephenson's *Snow Crash* (1992)
	Neal Stephenson's *The Diamond Age* (1995)
	Melissa Scott's *Night Sky Mine* (1997)
	Bruce Sterling's *Distraction* (1998)
	John Shirley's *Eclipse* (1999)
Transhumanism	Greg Egan's *Diaspora* (1997)
	Alastair Reynolds's *Revelation Space* (2000)
	Richard K. Morgan's *Altered Carbon* (2002)
	Greg Egan's Schild's *Ladder* (2002)
	David Mitchell's *Cloud Atlas* (2004)
	Peter Watts's *Blindsight* (2006)
	Liu Cixin's *The Three-Body Problem* (2007)
	Neal Stephenson's *Anathem* (2008)
	Suzanne Collins's *The Hunger Games* (2008)
	Andy Weir's *The Martian* (2011)
	David Brin's *Existence* (2012)
	Kim Stanley Robinson's *Aurora* (2015)
	Nnedi Okorafor's *Binti* trilogy (2015–2018)
	Martha Wells's *All Systems Red* (2017)

Science Fiction as Science Speculation

The building of a fictional world can provide a rich backdrop against which protagonists and antagonists can interact. The logic and structure of the fictional world can constrain and shape the interactions, providing opportunities for exploring new challenges and future realities. When discussing the characteristics of literary fictional worlds Doležel (1998) suggested that:

- fictional worlds are constructs of textual poiesis (through words things are brought into existence which did not exist before);
- fictional worlds are accessed through semiotic channels (the reader must cross the boundary between the realm of the actual and the realm of the possible);
- fictional worlds are ensembles of nonactualised possible states of affairs;
- fictional worlds are incomplete;
- the set of fictional worlds is unlimited and maximally varied;
- fictional worlds can be heterogeneous in their macrostructure (they can be made up of any number of differing components from different sources).

Doležel (1998) also suggested that fictional worlds inevitably contain logical gaps which require filling by the imagination of the reader. Some facts which are needed to understand the fictional world are provided directly in the text, while others are arrived at through deduction. Doležel used the term "saturation" to describe the amount of facts which are explicitly or implicitly provided by the author, and the term "gap" to describe those facts and behaviours which the reader must arrive at through imagination and extrapolation.

Controlling the size and nature of the factual gaps is often the defining characteristic of a work of science fiction. Questions faced by all science fiction authors are which real-world facts to maintain, which instead to extend and which fantastic new possibilities to introduce. And the manner of handling the introduction of the fantastic new possibilities is often a defining characteristic of the individual author.

For example, H.G. Wells is credited with having followed what is today referred to as Wells's Law which states that a fantasy story or work of science fiction should involve only a single extraordinary assumption. In the paragraphs below, maintained in his original words, Wells (1934) explains the approach.

> Anyone can invent human beings inside out or worlds like dumbbells or a gravitation that repels. The thing that makes such

imaginations interesting is their translation into commonplace terms and a rigid exclusion of other marvels from the story. Then it becomes human. "How would you feel and what might not happen to you", is the typical question, if for instance pigs could fly and one came rocketing over a hedge at you. How would you feel and what might not happen to you if suddenly you were changed into an ass and couldn't tell anyone about it? Or if you became invisible? But no one would think twice about the answer if hedges and houses also began to fly, or if people changed into lions, tigers, cats and dogs left and right, or if everyone would vanish anyhow. Nothing remains interesting, where anything may happen.

For the writer of fantastic stories to help the reader to play the game properly, he must help him in every possible unobtrusive way to domesticate the impossible hypothesis. He must trick him into an unwary concession to some plausible assumption and get on with his story while the illusion holds …

As soon as the magic trick has been done the whole business of the fantasy writer is to keep everything else human and real. Touches of prosaic detail are imperative and a rigorous adherence to the hypothesis. Any extra fantasy outside the cardinal assumption immediately gives a touch of irresponsible silliness to the invention.

Introducing a small number of new or unusual scientific facts, then exploring the consequences, has been the basis of much science fiction writing. Many stories have explored the nature of physical reality and the characteristics of the human condition, one small step at a time. And due in part to the incremental approach, many innovations and many new societal tendencies have been described first in science fiction then only later in science fact.

The manner of handling the new science is a criteria which is often cited in science fiction circles. A distinction is made between "hard" and "soft". The term "hard science fiction" is usually used to describe stories which strongly leverage scientific principles from fields such as chemistry, computer science, engineering, mathematics or physics. The term "soft science fiction" usually refers instead to those stories which leverage more economics, history, psychology, sociology or politics in their storylines. Hardness usually implies a rather close correspondence to known scientific facts about the real world we live in, while softness is often taken as shorthand for stories characterised by unknown or untested science.

Taking his inspiration from German mineralogist Friedrich Mohs who had developed a scale to rate minerals from 1 (softest) to 10 (hardest),

Sterling (2017) proposed a conceptually similar approach for placing works of science fiction on a spectrum from soft to hard. The "Mohs Scale of Science Fiction Hardness" (see Table 4.2) defined by Stirling describes the logical possibilities starting from a value of 1 for the most imaginative works of fantasy which are little anchored by known facts, and ending at a value of 6 for works which are mostly grounded in known real world scientific facts.

TABLE 4.2 The Mohs Scale of Science Fiction Hardness.

1	Science in Genre Only: the story is unambiguously set in the literary genre of science fiction but is not scientific. Applied phlebotinum is the rule of the day, often of the nonsensoleum kind. Green rocks gain new powers as the plot demands, and both Bellisario's Maxim and the MST3K Mantra apply.
	Futurama, *Star Wars*, *Tengen Toppa Gurren Lagann*, the DC and Marvel universes, *Doctor Who* and *The Hitchhiker's Guide to the Galaxy* are in this category.
2	World of Phlebotinum: the universe is full of applied phlebotinum with more to be found behind every star, but the phlebotinum is dealt with in a fairly consistent fashion despite its lack of correspondence with reality. Within the suggested universe the phlebotinum considered to lie within the realm of scientific enquiry.
	Neon Genesis Evangelion, the *Star Trek* series and *StarCraft* are in this category.
	This class includes a subclass (let's say 2.5 on the scale) which contains stories that are generally sound, except that the physics is not earth physics. These stories are often a philosophical exploration of a concept no longer considered true (such as Aristotelian physics or the Luminiferous Ether) or never considered true in the first place (e.g. two spatial dimensions instead of three, like Flatland).
	Some Arthur C. Clarke stories are in this category.
3	Physics Plus: stories that still involve multiple forms of applied phlebotinum but where the author justifies the creations with natural laws, both real and invented. And the creations from the same laws turn up again and again in new contexts.
	Schlock Mercenary, David Weber's *Honor Harrington* series, David Brin's *Uplift* series and *Battlestar Galactica* are in this category.
4	One Big Lie: the author invents one (or, at most, a very few) counterfactual physical laws and writes a story that explores the implications. Consider, for instance, Cities in Flight's "Dirac

(Continued)

TABLE 4.2 Cont.

Equations" and "spindizzy motor" leading to instantaneous communication. Or Mass Effect's "Element Zero" being the basis for all of the series' futuristic technology.

Most works in Alan Dean Foster's *Humanx Commonwealth* series, the Ad Astra board games, Robert A. Heinlein's *Farnham's Freehold* and many of Vernor Vinge's books are in this category.

This class includes a subclass (4.5 on the scale) which could be called One Small Fib. It contains stories that involve only a single counterfactual device (often Faster Than Light Travel) which is not a major element of the plot.

Mission of Gravity, Close to Critical and *Freefall* are in this category.

5   Speculative Science: stories in which there is no "big lie". The science is genuine speculative science or engineering, and the goal of the author is to make as few errors as possible with respect to known facts.

The first two books in Robert L. Forward's *Rocheworld* series and Robert A. Heinlein's *The Moon Is a Harsh Mistress* are in this category.

This class includes a subclass (5.5 on the scale) of Futurology. Stories which function almost like a prediction of the future, extrapolating from current technology rather than inventing major new technologies or discoveries.

Gattaca, Planetes, The Martian, Transhuman Space and the more speculative works of Jules Verne are in this category.

6   Fiction in Name Only: non-fiction stories where the science of our universe serves as the basis of the story, even if overly emphasised or exaggerated.

The *Apollo Program, World War II* and *Woodstock* are in this category.

Source: Adapted from Sterling (2017)

It is worth noting that the distinction between hard science fiction and soft science fiction is not always clear-cut. Many stories incorporate elements of both, and individual literary critics may have different criteria for what constitutes hard or soft. Nevertheless, the scale can prove useful in helping to highlight how much known real-world science the fictional world contains. And towards clarifying the focussing potential and deductive power of the fictional world. It helps to clarify whether or not the fictional world of the story comes close to being a possible world.

Stockwell (2000) has suggested that:

> The notion of possible world was developed in philosophical logic to resolve a number of problems to do with determining the truth or

falsity of propositions. The basic premise of all possible worlds theories is that our world – the actual world – is only one of a multitude of possible worlds.

A possible world is a hypothetical scenario which provides a context within which a given concept can be checked for compatibility and consistency. Possible worlds are often used in philosophy and physics to establish what is possible, contingent or necessary. Or simply to explore the implications of a particular concept. Possible worlds provide a basis for performing thought experiments on complex concepts and moral dilemmas. And, unlike a scientific experiment which involves the collection of external empirical data, a thought experiment is conducted entirely within the mind of the individual.

Thus if a fictional world of a story comes close to being a possible world, usually through "hardness", then the plot and character dynamics of the story can be considered indicative of what might happen in similar circumstances in the real world in the future. The story can be considered to have predictive power in relation to future developments.

And with predictive power comes influence. If a story seems realistic enough to predict a possible future development, it is likely to help towards making that future development happen. Von Stackelberg and McDowell (2015) have, for example, suggested that:

> Narratives about the future can trigger new directions for thought and exploration that foster the creation of new realities. The self-lacing shoes from *Back to the Future II*, a science fiction film released in 1989, became a reality in 2015 when Nike's innovation chief, who designed the shoes for the film, announced plans to release the shoes as a commercial product.

Another example of the facilitation is the film *Minority Report* (2002) which has been claimed to have facilitated the development of gesture-driven computer interfaces, facial recognition technologies, personalised advertising, driverless cars and robotic insects. And many other examples can be cited of products, systems or services which resemble their science fiction precedents. And which likely benefited from the science fiction explorations as proof of concept.

It has often been claimed that science fiction stories stimulate, or at least facilitate, innovation. Dourish and Bell (2014) suggested that

> ... science fiction in popular culture provides a context in which new technological developments are understood. Science fiction visions

appear as prototypes for future technological environments – the visualizations of photo enhancement and search technology in Ridley Scott's (1982) *Blade Runner* for instance presages contemporary digital image manipulation technologies by nearly two decades.

And Bleecker (2009) has claimed that:

The science fiction film is arguably much more effective than the more generally understood way of creating and sharing scientific knowledge, peer review protocols notwithstanding. The film adds a kind of idea-mass to something like Ubicomp that spreads the story much further, gives it more meaning in the context of a reasonable, non-ideal, flawed, human social world and does so with more momentum than a dowdy paper in an obscure, difficult to find science journal that, at best, comes up with scenarios that are about as realistic as a laptop that never crashes or wireless phone networks that never drop calls. The Ubicomp science fiction can bring to light consequences, conclusions and implications much better than a science fact paper, or awkward laboratory demonstration.

And Shedroff and Noessel (2012) have noted science fiction's influence on design with:

Why look to fiction for design lessons at all? How can it inform our non-fictional, real-world design efforts? One answer is that, whether we like it or not, the fictional technology seen in sci-fi sets audience expectations for what exciting things are coming next. A primary example is the *Star Trek* communicator, which set expectations about mobile telephony in the late 1960s, when the audience's paradigm was still a combination of walkie-talkie and the Princess phone tethered to a wall by a chord ... Exactly 30 years later, Motorola released the first phone that consumers could flip open in the same way the Enterprise's officers did.

Science fiction stories appear to have facilitated many technological developments. A detailed analysis by Shedroff and Noessel (2012) has suggested eight areas where science fiction stories have been particularly helpful:

- anthropomorphism;
- mechanical controls;
- visual interfaces;

- sonic interfaces;
- gestural interfaces;
- volumetric projection;
- augmented reality;
- brain interfaces.

In each of these areas there are precedents in science fiction stories, sometimes going back more than a century, which have opened the conversation, explored the issues and led the way for the real world developments.

And beyond facilitating innovation there is little doubt that many science fiction authors have also used their stories to highlight an issue or make a point. The fictional world, often hard enough to be considered a possible world, immersed the reader in a physical or moral dilemma deeply enough to produce emotion. Many stories have challenged readers to experience a physical or moral dilemma rather than just think about it.

And the approach to highlighting the issue or making the point has varied. Von Stackelberg and McDowell (2015) have suggested that future-oriented science fiction stories tend to fall into one of four literary categories:

- cautionary tales that emphasise the consequences, generally negative, of some aspect of today's society;
- thought experiments which are also referred to as "what if" stories, that examine the potential impacts of some current or anticipated event, technology, or trend;
- literalised metaphors that use a metaphor to study a particular aspect of our world and make it concrete, for example stories of space aliens to address our alienation from society;
- explorations of new science and technology that use new advances as the basis for a storyline.

Such observations suggest that science fiction authors have structured their individual stories as either a warning, a scenario, an association or a set of observations. Each of the approaches being capable of producing emotions and of challenging the reader to experience rather than just think.

It was not the intention of the author to provide a detailed history of the science fiction genre because the topic is engagingly elaborated elsewhere (see, for example, Roberts 2000; Seed 2011; Parrinder 2014). The material provided here is intended only as a short introduction to the typical themes, tropes and techniques of the genre. From the early works of Verne, Wells, Asimov, Heinlein and others to the more recent contributions of Dick, Gibson, Sterling, Cixin and many others, science fiction has provided a wealth of ideas and insights. In what remains of this chapter a short summary will be presented of what the world of science fiction has had to say about autonomous road vehicles, setting the scene for the later chapters of this book.

Science Fiction Speculations about Road Vehicles

While not as prevalent as rockets or spaceships, many human-driven or autonomous road vehicles can be found among the pages of science fiction. Perhaps less impactful on the imagination than some of the other forms of transport, road vehicles are nevertheless to be found in many stories. And they often exhibit unusual capabilities which reveal important aspects of the story's characters and plot.

And from the early 20th century the stories have explored many of the automobility themes which are familiar to us today. Concerns such as insufficient parking, traffic congestion, noise pollution, air pollution, dangerous drivers, road safety, autonomous vehicle malfunction, corporate power and human dependency are all to be found already among Golden Age stories (Withers 2020).

For example, in David H. Keller's 1928 story *The Revolt of the Pedestrians* the problems of noise, air pollution and vehicle dependency were explored. In the story the human race evolves into two species, one called the "automobilists" who live their lives tethered to mini cars which they use for all physical movement, and another called the "pedestrians" who reject vehicles and insist on moving under their own power. Pedestrians are ruthlessly persecuted and eventually forced to live in isolated enclaves until, one day, they find a way of halting the atomic energy machinery which permits the vehicles to move. With the machinery halted, the automobilists are left to starve in their automobile-dependent cities.

And in George McLociard's 1929 story *The Terror of the Streets* the phenomenon of dangerous drivers is exposed and the dangers of speed are explored. Shocked and saddened by the many road deaths, Professor Stefenson sets out to build a powerful machine called the TERROR which he uses to scare others into driving slowly and safely. Being too fast and powerful for the authorities to catch, the TERROR rampages through the city while newspapers warn residents to drive carefully because the TERROR may come at them out of nowhere.

As soon as road vehicles became common on roads they also became useful antagonists, protagonists, tools and metaphors for exploring issues and for stimulating human emotions. Over the years many science fiction stories have used a road vehicle to present an opportunity or a danger, and many others have used them as a metaphor for technology or for human progress in general.

A few well-known road vehicles from the world of science fiction which the reader will probably be familiar with are those shown in Figure 4.1.

Each is iconic and memorable and played an important role in the respective TV show or movie. And the aesthetics, dynamics, language capabilities and personality of each provided important clues to the nature of the fictional world and to the plot of the story.

Figure 4.1 Examples of road vehicles from the world of science fiction (from left to right, top to bottom): Batmobile from TV series *Batman* (1966–1968), Los Angeles police car from *Bladerunner* (1982), KITT from *Knight Rider* (1982–1986), DeLorean DMC-12 from *Back to the Future* (1985), M577 Armoured Personnel Carrier from *Aliens* (1986), Batmobile from *Batman* (1989), taxi from *Total Recall* (1990), taxi from *The Fifth Element* (1997), Lexus Maglev Pod 2054 from *Minority Report* (2002) and Audi RSQ from *I, Robot* (2004).

Source: Max Sims

Each had an advanced propulsion system to move faster and more efficiently than other vehicles as it transported the main character or characters. In each case the vehicle's equipment and behaviour helped to reveal character traits of the story's antagonists and protagonists.

Several of the vehicles had special items of equipment such as the Batmobile's gadgets, KITT's surveillance systems and the M577's heavy armour and weaponry for military operations. The DeLorean DMC-12 from *Back to the Future* even had technology for travelling through time.

And most of them were equipped with artificial intelligence systems which assisted their drivers with communication and navigation. For example the 1989 Batmobile had a built-in communication system that allowed it to contact the Batcave and KITT had advanced navigation and mapping capabilities. Several were even capable of operating autonomously, the Los Angeles police car from *Bladerunner*, KITT from *Knight Rider* and the autonomous taxis of *Total Recall* being obvious examples.

Considering specifically autonomous road vehicles, the first example is often considered to be the automated delivery vehicle in Miles J. Breuer's short story *Mechanocracy*. Published in 1932 in *Amazing Stories*, *Mechanocracy* is an early example of a future society which is controlled by machines. Most of the decision-making is done by machines and the world is coordinated by a mechanical brain called the Centralized Authority. The story includes an automated delivery vehicle whose travel destination is selectable via switches from a small predetermined list. In Breuer's words:

> Out there was a row of trucks with laundry packages dropping into them from overhead chutes. They were automatic trucks such as are used for making deliveries beyond the pneumatic-tube zone. They leaped into the foremost truck. Quentin set the switches on impulse for Bay Shore, because that was not where he wanted to go, and they both rolled back into the closed portion of the vehicle. The truck started slowly, gathered momentum, and automatically made its way out of the city. After fifteen minutes of eternity, they looked out. The truck was moving swiftly along a country road.

Already at this early stage there are suggestions of the simplifications which can be achieved through the use of fixed predetermined stops such as the "Bay Shore" of the story. And already at this early stage there can be found hints of autonomous road vehicles filling the gaps in the travel network, such as the "pneumatic tube zone" of the story.

And autonomous road vehicles were the main focus of David H. Keller's 1935 story *The Living Machines*. Horrified by the many road fatalities, an inventor named Poorson develops a self-driving automobile controlled by a mysterious sphere. The story explores the benefits of self-driving vehicles for people with disabilities, shy individuals, children and other members of the community who were previously denied access to transport. But the story also explores the dangers. It concludes with an episode of industrial sabotage where the gasoline supply is contaminated with cocaine causing the machines to vandalise, terrorise and murder.

Robert Heinlein's 1951 book *Between Planets* provides one of the earliest examples of a robo-taxi. It mentions the control systems needed for robo-taxis to function and alludes to the benefits which such systems provide when navigating the streets of a large city. One passage describes the destination selection and the efficiency of navigation. In Heinlein's words:

> He dialled the address and settled back. The little car lurched forward, climbed a ramp, threaded through a freight tunnel, and mounted an elevator. At first Don tried to keep track of where it was taking him but the tortured convolutions of the ant hill called "New Chicago" would have made a topologist dyspeptic; he gave up. The robot cab seemed to know where it was going and, no doubt, the master machine from which it received its signals knew.

Illustrative instead of New Wave sensitivities and of the growing descriptive sophistication, Alan Dean Foster's 2006 book *Sagramanda* highlighted the difficulties which robo-taxis face when balancing customer requests against the constraints of road rules and city ordinances. In Foster's words:

> he urgently addressed the vehicle's AI. "Can't we go any faster? I'm already running late".
>
> Since the taxi utilized sophisticated electronic sensors to perceive its surroundings, the traditional forward windshield existed only to allow fares to see where they were going. The vehicle was as aware of this as its passenger.
>
> "As you can see, sir, this is a very busy street, and I am forbidden by law and by coding from forcing a path …"
>
> All were equipped with the same city-regulated programming.

And the need for robo-taxis and other autonomous road vehicles to have some form of human backup when encountering new or unusual situations is also a theme which is encountered in several science fiction stories. For example, Philip K. Dick's 1955 short story *Autofac* explored the concept of robots taking over human jobs and described a situation where human assistance was requested by an autonomous road vehicle. In Dick's words:

> The truck was massive, rumbling under its tightly packed load. In many ways, it resembled conventional human-operated transportation vehicles, but with one exception – there was no driver's cabin. The horizontal surface was a loading stage, and the part that would normally be the headlights and radiator grill was a fibrous spongelike mass of receptors, the limited sensory apparatus of this mobile utility extension.
>
> Aware of the three men, the truck slowed to a halt, shifted gears and pulled on its emergency brake. A moment passed as relays moved into action; then a portion of the loading surface tilted and a cascade of heavy cartons spilled down onto the roadway. With the objects fluttered a detailed inventory sheet.
>
> The truck had begun to move away; abruptly it stopped and backed toward them. Its receptors had taken in the fact that the three men had demolished the dropped-off portion of the load. It spun in a grinding half circle and came around to face its receptor bank in their direction. Up went its antenna; it had begun communicating with the factory. Instructions were on the way.

A particularly illuminating example of an autonomous road vehicle in science fiction is that of the automatobile "Sally" of Isaac Asimov's 1953 short story of the same name. The story leveraged several themes and tropes which have now become familiar to readers. For example, the story includes an early description of the use of autonomous road vehicles in taxi fleets. In Asimov's words:

> Of course, the automatics were ten to a hundred times as expensive as the hand-driven ones, and there weren't many that could afford a private vehicle. The industry specialized in turning out omnibus-automatics. You could always call a company and have one stop at your door in a matter of minutes and take you where you wanted to go. Usually, you had to drive with others who were going your way, but what's wrong with that?

And the story alludes to the issues of legislation and public acceptance which are so typical in the autonomous road vehicle discourse of today. In Asimov's words:

We take it all for granted now, but I can remember the days when the first laws came out forcing the old machines off the highways and limiting travel to automatics. Lord, what a fuzz. They called it everything from communism to fascism. But it emptied the highways and stopped the killing, and still more people get around more easily the new way.

Asimov's short story *Sally* conveys a warning which is only superficially disguised as narrative, that of the dangers of autonomous road vehicles being given excessive agency, or, worse, becoming sentient and adopting plans of their own.

Throughout the story the adopted trope is that of the creation which escapes the control of its creator, alluding to the already then present societal concerns in relation to the ethics, agency and sentience of machines. Asimov's views about the risks of excess agency are bluntly expressed at the end of the *Sally* short story in the protagonist's worried reflection:

> There are millions of automatobiles on Earth, tens of millions. If the thought gets rooted in them that they're slaves; that they should do something about it … If they begin to think the way Gellhorn's bus did … Maybe it won't be till after my time. And then they'll have to keep a few of us to take care of them, won't they? They wouldn't kill us all. And maybe they would. Maybe they wouldn't understand about how someone would have to care for them. Maybe they won't wait. Every morning I wake up and think, Maybe today …

Sally and her science fiction brethren have served as testbeds for autonomous road vehicles. Fictional machines in fictional worlds which have suggested both the potential and the dangers. Many, perhaps most, of the key concepts which can be found in today's autonomous road vehicle discourse can also be found in bits and pieces among the pages of science fiction. And many, perhaps most, of the key challenges for autonomous road vehicle designers were also implicitly or explicitly raised. So what lessons can be learned about autonomous road vehicles from the world of science fiction?

One observation is Roberts's claim (2000) that

> … in most cases technology works in science fiction either directly or obliquely to collapse together the machine and the organic. The bulk of Science Fiction technology articulates the trope of the

Cyborg, the machine/organic hybrid that is both a special instance of technology and the emblem for all of it.

Roberts's claim focusses attention on the overlapping spheres of capability and responsibility which characterise the relationship. Indeed, much of today's autonomous road vehicle discourse focusses on the difficulty of integrating people and machines and highlights the safety risks and the social repercussions involved.

Another observation is that Sally and her science fiction brethren have shown us the confusion which exists in society in relation to autonomy. As Murphy (2019) has suggested: "The term 'autonomy' has so many societal connotations that it interferes with an informed understanding of artificial intelligence. The media often use autonomy to connote a robot with human-equivalent intelligence and sentience that can also act with unbounded initiative". Adding that "... artificial intelligence re-searchers and roboticists think of autonomy differently than what is presented in the media".

Murphy (2019) identified six areas of divergence between societal conceptions of autonomy as manifested in the science fiction literature and the actual work of roboticists:

- robot autonomy is not the same as intelligence;
- robot autonomy is not the same as sentience;
- robot autonomy is not the same as initiative;
- robot autonomy is not the same as political autonomy;
- machine learning is not the silver bullet that will produce constraint autonomy;
- lack of trust of autonomy is believed to be primarily due to the association of autonomy with initiative combined with inappropriate testing and evaluation methods.

Murphy exemplified the differences by considering the case of Isaac Asimov's influential novel *I, Robot*. Table 4.3, based on Murphy's analysis, suggests the novel's selective focus on anthropomorphic robots and on human-like behaviours to the exclusion of robots which serve their intended purposes differently. Despite *I, Robot* containing several different short stories involving several different robots in several different contexts, all of the robots exhibit anthropomorphic features, provide somewhat similar benefits and present similar dangers.

To facilitate reader engagement the autonomy which is found in works of science fiction is usually extensive and is usually accompanied by human-like behaviour. Works of science fiction have tended to

TABLE 4.3 Robot characteristics in Asimov's *I, Robot.*

	Examples in *I, Robot*	Lack of Examples in *I, Robot*
Morphology	humanoid	non-anthropomorphic
Autonomy	full	shared
Intelligence	artificial general intelligence	narrow specialised intelligence
Communication	naturalistic	physical, graphical or brain–machine interface

Source: Adapted from Murphy (2019)

increase the immediacy of reader interpretation by deploying robots which are more similar to humans than what is likely. The empathy, joys and fears of the readership have been stimulated by machines which are probably rather different from those which will find their way onto our streets.

The scarcity of narrowly intelligent, non-humanoid robots in the science fiction literature is unfortunate. Even today there are still few examples to be found of machines which lie between the single function or remotely controlled drone on the one hand, and the highly capable intelligent humanoid robot on the other. The pursuit of relatedness and of narrative clarity seems to have limited the variety, and thus the opportunities for exploring differently capable and differently intelligent machines.

This scarcity is perhaps one of the greatest limitations of the genre because it has curbed the exploration of those many possible forms of agency which are not humanoid in nature. There have been few examples of robots in science fiction which express their individuality, interactional asymmetry and normativity in manners significantly different from humans. This is unfortunate given the science fiction author Larry Niven's (2002) notion that "The only universal message in science fiction: there exist minds that think as well as you do, but differently".

And when considering autonomous road vehicles Murphy (2020) has noted that:

> What science fiction as a whole missed, with the exception of Imperial Earth, were the legal and ethical implications of autonomous cars. The legal questions as to product liability and safety testing and evaluating cars on public streets are never addressed. Novels and stories introduce autonomous cars as being everywhere

while avoiding how such technologies became ubiquitous. Science fiction has also ignored ethics. For example, KITT has features that, if incorporated into personal cars, could prevent driving deaths, but the inventor restricts the technology to a single individual performing vigilante crime fighting, implying that this is the greatest public good. This is rather like developing a cure for cancer but giving it to a single police officer rather than distributing it to the general public.

Murphy concluded his analysis with:

the value of autonomous cars for saving lives is unquestionable, but the conversation is still about what society should do to prevent a potential artificial intelligence uprising. More useful conversations might be about making sure everyone benefits from safety advances as soon as possible and how to find the right balance of regulations, capitalism, and research that will ensure safe experimentation while sustaining innovation.

Sally and her science fiction brethren have provided engaging and thought-provoking entertainment, raising many important questions in relation to autonomous road vehicles along the way. Nevertheless, the focus of the stories has tended to be on matters which are not necessarily the most pressing. Suspenseful and entertaining plots have been achieved based on far-future characteristics such sentience and intellectual superiority, but such concerns may not be among the most immediate or even among the most problematic.

This chapter has provided an introduction to what is meant by the term "science fiction" and has briefly reviewed the history of the genre. Major themes, tropes and techniques were noted. The natures of world building, fictional worlds and possible worlds were discussed as was the use of possible worlds as preliminary testbeds for new technologies and new societal trends. Historical examples were cited in support of the proposal that science fiction stories can facilitate new developments by acting as forms of science speculation.

This chapter has also reviewed a few of the road vehicles, particularly the autonomous road vehicles, which are found among the pages of science fiction stories. It was noted that road vehicles are found already in the earliest works and that many, possibly most, of the themes of current road vehicle discourse are present. And it was noted that possibly one of the greatest limitations of the genre has been its focus on highly autonomous and highly human-like robots. The pursuit of relatedness and of narrative clarity seems to have limited the variety, and thus the opportunities for exploring differently capable and differently intelligent machines. Narrowly intelligent, non-humanoid, robots are rare among the pages.

Having reviewed the world of science fiction stories, which are only partially constrained by real-world facts, the next chapter introduces instead a few of the more constrained tools which real-world designers use when speculating about the future. The next chapter briefly reviews the history of each tool, introduces its main characteristics and notes its typical conditions of use. It provides a set of options which are available to designers for use when designing the future friendly neighbourhood robots. And it ends with a few clarifications about the tool which was chosen for the speculations which form the basis of the later chapters of this book.

References

Ad Astra Institute for Science Fiction & the Speculative Imagination 2022, https://adastra-sf.com/SF-Defined.htm#SF-defined.

Asimov, I. 1950, *I, Robot*, The Gnome Press, New York, New York, USA.

Asimov, I. 1969, Sally, in *Nightfall and Other Stories*, Fawcett Publications, Greenwich, Connecticut, USA.

Bleecker, J. 2009, *Design Fiction: a short essay on design, science, fact and fiction*, Near Future Laboratory, USA.

Breuer, M.J. 1932, Mechanocracy, *Amazing Stories*, Experimenter Publishing, Vol. 7, No. 1, April, New York, New York, USA.

Dick, P.K. 1955, Autofac, *Galaxy Science Fiction Magazine*, Galaxy Publishing Corporation, Vol. 11, No. 2, November, Boston, Massachusetts, USA.

Doležel, L. 1998, *Heterocosmica: fiction and possible worlds*, The Johns Hopkins University Press, Baltimore, Maryland, USA.

Dourish, P. and Bell, G. 2014, "Resistance is futile": reading science fiction alongside ubiquitous computing, *Personal and Ubiquitous Computing*, Vol. 18, no. 4, pp. 769–778.

Foster, A.D. 2006, *Sagramanda: a novel of near future India*, Pyr Publishing, Amherst, New York, New York, USA.

Goodwin, M.A. 2020, *Neo Cyberpunk – the anthology*, independently published via Amazon Kindle Direct Publishing.

Heinlein, R.A. 1951, *Between Planets*, Charles Scribner's Sons, New York, New York, USA.

Huxley, J. 1957, *New Bottles for New Wine*, Chatto & Windus, London, UK.

Keller, D.H. 1928, The revolt of the pedestrians, *Amazing Stories*, February, Experimenter Publishing, New York, USA.

Keller, D.H. 1935, The living machines, *Wonder Stories*, May, Stellar Publishing, New York, USA.

Kingsley, A. 1961, *New Maps of Hell: a survey of science fiction*, Victor Gollancz, London, UK.

McLociard, G. 1929, The terror of the streets, *Amazing Stories*, April, Experimenter Publishing, New York, USA.

Murphy, R.R. 2019, *Learn AI and Human–Robot Interaction from Asimov's I, Robot Stories, Robotics through Science Fiction*, Vol. 2, Robin R. Murphy, printed by Amazon, UK.

Murphy, R.R. 2020, Autonomous cars in science fiction, *Science Robotics*, Vol. 5, No. 39, eaax1737.

Niven, L. 2002, Niven's Laws, *Analog: Science Fiction and Fact*, November issue, Dell Magazines, Norwalk, Connecticut, USA.

Parrinder, P. (ed.) 2014, *Science Fiction: a critical guide*, Routledge, London, UK.

Person, L. 1998, Notes toward a postcyberpunk manifesto, *Nova Express*, Vol. 4, No. 4, Whole Number 16, Winter/Spring.

Roberts, A. 2000, *Science Fiction*, Routledge, London, UK.

Seed, D. 2011, *Science Fiction: a very short introduction*, Oxford University Press, Oxford, UK.

Shedroff, N. and Noessel, C. 2012, *Make It So: interaction design lessons from science fiction*, Rosenfeld Media, Brooklyn, New York, USA.

Sterling, B. 1986, Preface to *Burning Chrome* by William Gibson, Gollancz, London.

Sterling, B. 2017, The Mohs Scale of Science Fiction Hardness, *Wired*, 25 November.

Stockwell, P. 2000, *The Poetics of Science Fiction*, Routledge, London.

Suvin, D 1979, *Metamorphoses of Science Fiction: on the poetics and history of a literary genre*, Yale University Press, New Haven, Connecticut, USA.

Von Stackelberg, P. and McDowell, A. 2015, What in the world? Storyworlds, science fiction, and futures studies, *Journal of Futures Studies*, Vol. 20, No. 2, pp. 25–46.

Wells, H.G. 1934, *Seven Famous Novels*, Alfred A. Knopf Publisher, New York, New York, USA.

Withers, J. 2020, *Futuristic Cars and Space Bicycles: contesting the road in American science fiction*, Liverpool Science Fiction Texts and Studies Vol. 66, Liverpool University Press, Liverpool, UK.

Chapter 5

Speculative Approaches

Scenarios

One design approach for speculating about the future is that of "scenarios". Alcamo and Ribeiro (2001) have suggested that the word is borrowed from the world of performance theatre where it refers to the sequential elements of a screenplay, such as the actions of the performers or the changes in the stage setting. The term is widely used to refer to a sequence of events or an account of a projected course of action. It includes plausible descriptions of the actors involved, their motivations, the locality, the context and the technological, environmental and societal constraints. The use of scenarios is today a routine part of many design projects and designers often go to great lengths to ensure the validity and usefulness of their descriptions.

The earliest formal scenarios are usually considered to be those of Herman Kahn (Kahn and Wiener 1967) and the RAND Corporation which described the possible build-up to, execution of, and effects of nuclear war. Often claimed to be a founding father of the scenario approach, Kahn's scenarios were characterised by a focus on quantitative data and systematic analysis of the available options. They developed a reputation for being useful for structuring discussions and for achieving agreement between the parties.

From 1971 Pierre Wack is credited with having introduced the use of scenarios at Royal Dutch Shell, which the company put to intensive use during the oil shocks two years later. The arrival of the oil supply crisis of the mid-1970s saw the widespread adoption of planning scenarios based on hypothesis about the total amount of available oil in the world and the likely political constraints to its extraction and use. The period of the oil shocks saw extensive use of scenarios and the spread of the approach from the military and political realm to areas of industrial, economic and social planning.

DOI: 10.4324/9781032724232-5

The early scenarios of the 1960s and 1970s tended to focus on strategic aspects of what might happen. They usually described events or tendencies of societal or even planetary significance as part of the effort to comprehend and discuss the challenges. Gradually, however, the approach also began to be applied to less dramatic questions.

And with the spread of digital technologies, scenarios came to be used for capturing the intentional and interactive aspects of situations involving designed artefacts. As Rosson and Carroll (2002) have suggested, the use of scenarios shifted the design focus from the functional system specifications to the sequence of user decisions and interactions. Scenarios helped to shift attention from the artefact to the people who use the artefact.

An early advocate of scenarios in digital systems design was Alan Cooper (1998) who suggested that it should be "... a concise description of a persona using a software-based product to achieve a goal". Cooper emphasised the benefit of two types of scenario: "daily-use scenarios" and "necessary-use scenarios". Daily-use scenarios capture the actions which are performed most frequently. While necessary-use scenarios provide instead a description of the actions which must absolutely be performable, even if only occasionally. Cooper considered the daily-use scenarios to be the more important of the two types, due to the spotlighting on the characteristics which manifest themselves most frequently.

Carroll (2000) has suggested that "A scenario is a story with a setting, agents or actors who have goals or objectives, and a plot or sequence of actions and events". And noted (Carroll 1997) that the scenarios used in human–computer interaction (HCI) help to identify design requirements and enhance usability by bringing into focus people's intentions and behavioural patterns.

Over the years scenarios have proven to be a useful approach for understanding how people interact with products, systems and services in different contexts. And for formulating the associated user testing. They have been helpful at capturing user needs and desires, at facilitating the discussion of those needs and desires, and at checking that the proposed softwares or devices meet those needs and desires (Mulder and Yaar 2006). Scenarios provide useful focal points for design and provide needed documentary evidence of the design decision-making.

Alcamo and Ribeiro (2001) suggested that the main elements of any scenario are:

- Description of step-wise changes: the main element of a scenario is the portrayal of step-wise changes in the future state of society and the environment. These changes can be expressed, for example, in the form of a diagram, table, or even as a set of written phrases.

- Driving forces: the main factors or determinants that influence the step-wise changes described in a scenario. Values for these driving forces must be assumed by the scenario developers, or taken from other studies.
- Base year: the beginning year of the scenario. For quantitative scenarios, the base year is usually the most recent year in which adequate data are available to describe the starting point of the scenarios.
- Time horizon and time steps: the time horizon describes the most distant future year covered by a scenario. The number of time steps between the base year and time horizon of the scenarios are usually kept to a minimum because of the large analytical effort needed to describe each year.
- Storyline: a narrative description of the scenario which highlights its main features and the relationships between the scenario's driving forces and its main features. These storylines can be newly constructed for each new scenario study, or they can be taken from previous scenario exercises.

And as shown in Table 5.1 they also proposed three ways of categorising scenarios. They can be distinguished based on what they are trying to achieve (exploratory or anticipatory), based on how they are decided upon (deductive or inductive) or based on the type of descriptive materials involved (qualitative or quantitative).

A recent comprehensive review of the history and use of scenarios in multiple fields of endeavour has led Cha (2019) to propose an updated definition of the term. In her words, "A scenario is a description of a sequence of events and activities that occur within a specific context, and which can involve other agents such as intelligent technologies, people or animals". The definition emphasises the contextual nature of the description and highlights the agencies of the actors involved.

The contextual emphasis of the definition reflects the fact that any scenario takes place within a storyworld, either past, present or future, which is characterised by a location, technology base, culture and timeframe. Thus all scenarios involve storyworld specific constraints on the possible actions and implications. The updated definition alludes to the need for consistency in maintaining the associated internal relationships.

And the definition also reflects the growing importance of agency. Where past scenarios may have focussed mainly on the motivations, actions and implications of humans, a growing number of scenarios now

TABLE 5.1 The three ways of categorising scenarios suggested by Alcamo and Ribeiro (2001).

Exploratory vs Anticipatory	Exploratory scenarios are those that begin in the present and explore trends into the future. This comes close to the original meaning of the word 'scenario' in the sense that it is a sequence of emerging events. By contrast, anticipatory scenarios start with a prescribed vision of the future (either optimistic, pessimistic, or neutral) and then work backwards in time to visualise how this future could emerge.
Deductive vs Inductive	Deductive scenarios are derived from a framework which organises the big uncertainties or questions about the future into a logical form. First the framework is established, then scenarios are deduced from the framework. Inductive scenarios are derived from taking into consideration all data and ideas about the future. Scenarios are built step-wise, bottom-up. First all data are considered, then insights about the future are induced from this study of the data.
Qualitative vs Quantitative	Qualitative scenarios describe possible futures in the form of words or visual symbols rather than numerical estimates. They can take the shape of diagrams, phrases, or outlines, but more commonly they are made up of narrative texts, the so-called 'storylines' mentioned earlier.
	Quantitative scenarios provide numerical information in the form of tables and graphs. Their disadvantage is that the exactness of their numbers is sometimes taken as a sign that we know more about the future than we actually do. Another disadvantage is that quantitative scenarios are usually based on results of computer models, and these contain many implicit assumptions about the future. It has been argued that these models tend to represent only one point of view about how the future will unfold, and in this way produce scenarios that are unnecessarily narrow in view.

also reflect the agencies which society is assigning to members of the animal kingdom and to artificial agents. Needs and sensitivities are changing, thus so are the scenarios.

Cha (2019) has also combined the recommendations of several researchers including, but not limited to, Ogilvy and Schwartz (1996), Schwartz (1996), Pruitt and Grudin (2003), Van Notten (2005) and Alcamo (2007) to propose that developing a design scenario involves the following steps:

- Set Boundary Conditions: define the major structural terms of reference of the design activity such as any physical, manufacturing or cost barriers, the intended geographical coverage, the target population and the target time window.
- Identify Key Driving Forces: obtain the opinions of the main stakeholders through workshops, surveys, interviews or Delphi methods and summarise the driving forces which the stakeholders mentioned in a qualitative and/or quantitative manner.
- Develop Mini-Scenarios: define mini-scenarios which consist of detailed descriptions of the engagement and interaction with one of the aspects or consequences of one or more of the driving forces. A single persona (individual) is usually used in a given mini-scenario so as to ensure a single character focus based on a single lifestyle and single point of view.
- Write Full Storylines: extend and elaborate each mini-scenario based on the character, lifestyle choices, emotional responses and other human characteristics of the associated persona. A storyline can be considered an enriched and upgraded mini-scenario which is appropriate for a general audience in a manner not dissimilar to a screenplay outline. The storyline development is guided by narrative criteria such as "what is done", "where is it done", "by whom", "when", "by what means" and "in what way".

The fourth activity described by Cha (2019) is currently undergoing much development as writing techniques are adapted to meet the needs of different design sectors and of different designers. How to achieve narrative focus on the most important characters, lifestyle choices, emotional responses and other human characteristics is a subject of much debate. As are the optimal writing style, length of narrative, depth of treatment and use of sector specific tropes.

For example, there has been a growing use of scenarios which describe contexts, goals, actions and events which stimulate the human emotions (Cha 2019). People's emotional states are inevitably influenced by the interactions with a given product, system or service. In some contexts a strong emotional response can prove problematic or even dangerous, negatively affecting decision-making and safety. In others, a strong emotional response can be helpful due to increasing the human engagement and satisfaction. Many research studies have noted that human emotional states influence safety, efficiency, experience and purchasing decisions. Thus designers often now use one or more scenarios which are written in such a way as to lead to either strong positive emotions, or strong negative emotions, or both.

And beyond their use in describing actors, motivations, localities, contexts and constraints, scenarios can also be used in an ideological manner as a rallying point. As noted early on by Pierre Wack of Royal Dutch Shell (Wilkinson and Kupers 2014) scenarios involve stories and tend to adopt literary plot elements to situate and contextualise the information. Thus a strength of the scenarios approach is the ability to convey information via the most natural human form of communication, storytelling.

Manzini (2003) has discussed what he refers to as "design-orienting scenarios" which he describes as

> … a set of motivated, structured visions that aim to catalyse the energy of the various actors involved in the design process, generate a common vision, and hopefully cause their actions to converge in the same direction. They consists of three fundamental components: a vision, a motivation, and a strategy.

Design-orienting scenarios are detailed expositions of a general objective or general principle. Through reasoning and through examples they articulate in detail the issues involved and the advantages offered. According to Manzini any design-orienting scenario should focus strongly on the specific characteristics of plurality, feasibility/acceptability, micro-scale, visual expression and participation which characterise the proposal.

Design-orienting scenarios thus act more as signposts pointing in a desired direction than as examples of situations which need exploring. Where a traditional design scenario might be deployed to answer "what if" questions or "why" questions or "when" questions, design-orienting scenarios are better suited to addressing questions of "by what means" and "in what way". Such scenarios typically benefit from a greater familiarity with the relevant facts and a greater certainty about which are the most relevant.

A simple example of a scenario in support of automotive design is presented in Figure 5.1. The daily-use scenario describes a key step in

On a motorway a driver concentrates on the road ahead.

Just before an exit a dashboard mounted navigator alerts the driver and requests confirmation of the proposed route.

Confirmation is by pushing either the words "proceed to route" or "suggest new route".

Figure 5.1 Example of a daily-use scenario: Dashboard Mounted Navigator Confirmation.

Source: Max Sims

the interaction between a driver and a vehicle on-board system. Such scenarios help to explore the nature of the interaction, identify any sources of friction and note the likely human emotional responses. And such scenarios can stimulate the ideation of different designs and of new interaction concepts.

Prototypes

Another speculative design approach is that of "prototypes". Prototypes reduce the pressures on human creative thought processes by removing the need to simultaneously maintain a large array of characteristics and interactions in mind. They render a concept more concrete by filling in imaginative gaps via real-world detail, often bringing to light characteristics or implications which were not initially imagined. And they provide verification that the concepts and characteristics which were imagined based on logic and experience do in fact translate into real-world artefacts.

Defining and building prototypes has over the years become a routine part of most design projects. And, as with scenarios, designers often go to great lengths to ensure the accuracy, validity and usefulness of their prototypes. Dictionary entries for the word "prototype" usually list at least two concepts:

- the original or model on which something is based or formed;
- someone or something that serves to illustrate the typical qualities of a class.

Formal definitions thus suggest that a prototype can be an early sample, model or release of an artefact which is built to test a concept or a process. Or instead it can be an artefact which is sufficiently typical of some category as to be considered representative of that category. The word can thus be used to refer to something which is exploratory of a possible future, or instead to something which is typical and represent-ative of the past or present. The first meaning being of greater relevance to the materials presented in this book.

In design settings a prototype is usually an early version of a product, system or service which is built to test and refine the concept before arriving at a final version. It can be virtual or physical, and may be created using a variety of tools and materials depending on the nature of the project. In design settings a prototype is often created rapidly and with a low level of detail in order to quickly evaluate and iterate. And it can be incomplete or imperfect, not having all of the features and details of the final version. A prototype is usually used to gather feedback from stakeholders and to evaluate the feasibility, usability and effectiveness of the concept.

Perhaps the most common understanding of the term "prototype" is that expressed by Houde and Hill (1997) who suggested that "We define prototype as any representation of a design idea, regardless of medium". And perhaps the most common reason for using a prototype is that described by Kirby (2010) who wrote:

Any number of obstacles can impede or alter the development of a potential technology including a lack of funding, public apathy over the need for the technology, public concerns about potential applications, or a fundamental belief that the technology will not work. For scientists and engineers, the best way to jump-start technical development is to produce a working physical prototype.

Suchman et al. (2002) have emphasised that prototypes are never self-sufficient, stand-alone artefacts. The thing in itself is never the complete story. Prototypes are always, to some extent, a social construct and a social facilitator. A point which emerges from Suchman's definition of prototype as "… an exploratory technology designed to effect alignment between the multiple interests and working practices of technology research and development, and sites of technologies-in-use".

In design practice the use of paper, plastic, electrical components and software code is widespread when prototyping. Certain materials and processes are convenient and help to contain costs. And computer-aided design and other virtual representations are increasingly popular. There are thus a variety of ways to prototype a product, system or service to test its concept.

And the choice of the best prototype to use is complicated by the fact that different prototypes involving different physical or virtual character-istics often produce roughly the same results. For example, Sefelin et al. (2003) reported that the use of paper prototypes or instead low-fidelity computer implementations led to almost the same quantity and quality of critical user statements and to the same overall conclusions in their study of computer interfaces. There are thus no hard-and-fast rules for choosing the best prototype to build and to use.

One distinction which is sometimes made in the world of prototyping is that between low fidelity and high fidelity. Rudd et al. (1996) have suggested that: "Low-fidelity prototypes are generally limited function, limited interaction prototyping efforts. They are constructed to depict concepts, design alternatives, and screen layouts, rather than to model the user interaction with a system". Whereas the same authors described high-fidelity prototypes as "…fully interactive. Users can enter data in entry fields, respond to messages, select icons to open windows and, in general, interact with the user interface as though it were a real product". The general characteristics of the two categories are listed in Table 5.2.

Rudd et al. further noted that low-fidelity prototypes are usually simple representations which are intended for checking the look and perhaps the feel of an artefact, rather than testing how it operates. And that low-fidelity prototypes often require a facilitator to explain or demonstrate the prototype to participants and stakeholders.

Instead, they considered high-fidelity prototypes to be more complex and costly implementations which more closely represent the definitive artefact. And which are sometimes sufficiently realistic as to be indistinguishable from it. High-fidelity prototypes trade off some func-tionalities and some realism in order to reduce development times and development costs but are nevertheless close, or at least closer, to the definitive artefact.

TABLE 5.2 Advantages and disadvantages of low-fidelity and high-fidelity digital prototypes.

Type	Advantages	Disadvantages
Low-Fidelity Prototype	Lower development cost. Evaluate multiple design concepts. Useful communication device. Address screen layout issues. Useful for identifying market requirements. Proof of concept.	Limited error checking. Poor detailed specification to code to. Facilitator-driven. Limited utility after requirements established. Limited usefulness for usability tests. Navigational and flow limitations.
High-Fidelity Prototype	Complete functionality. Fully interactive. User-driven. Clearly defines navigational scheme. Use for exploration and test. Look and feel of final product. Serves as a living specification. Marketing and sales tool.	More expensive to develop Time-consuming to create. Inefficient for proof-of-concept designs. Not effective for requirements gathering.

Source: Adapted from Rudd et al. (1996)

Houde and Hill (1997) have suggested that a prototype can be framed in terms of three dimensions:

- role;
- look and feel;
- implementation.

And suggested that:

Each dimension corresponds to a class of questions which are salient to the design of any interactive system. "Role" refers to questions about the function that an artifact serves in a user's life – the way in which it is useful to them. "Look and feel" denotes questions about the concrete sensory experiences of using an artifact – what the user looks at, feels, and hears while using it. "Implementation" refers to questions about the techniques and components through which an artifact performs its function – the "nuts and bolts" of how it actually works.

There is no scarcity of theories in relation to prototyping but in practice the choice of the best one to use is usually dictated by what most needs checking. The best prototype is the one which best captures the issues to discuss and which best supports the decisions which must be made. The questions drive the characteristics. Cost and rapidly of fabrication are important considerations, but the main drivers of prototype selection are usually questions such as what might go wrong, is usage error-free or how pleasant and entertaining is the usage experience.

And like the prototypes themselves the motivations for developing them are varied. For example, Floyd (1984) has suggested three common reasons for prototyping:

- prototyping for exploration: the emphasis is on clarifying the requirements and desirable features for the target system and on discussing alternative solutions;
- prototyping for experimentation: the emphasis is on determining the adequacy of a proposed solution before investing in large-scale implementation of the target system;
- prototyping for evolution: the emphasis is on adapting the system gradually to changing requirements which cannot be reliably determined at an early stage of the design process.

And Lim et al. (2008) have suggested three guiding principles which can be considered when developing any prototype:

- Anatomy of prototypes: prototypes are filters that traverse a design space.
- Fundamental prototyping principle: prototyping is an activity with the purpose of creating a manifestation that, in its simplest form, filters the qualities in which designers are interested, without distorting the understanding of the whole.
- Economic principle of prototyping: the best prototype is one that, in the simplest and the most efficient way, makes the possibilities and limitations of a design idea visible and measurable.

Prototypes are thus artefacts which can have several uses and which can serve several purposes. As points of reference and centres of focus, they can serve to structure discussions and record decisions. As concrete manifestations of a concept, they can help detect issues and test hypothesis. And when reasonably close to the intended final artefact in terms of aesthetics and function, prototypes can be used in validation and marketing.

Thomas Edison is sometimes suggested to have said that "if a picture is worth a thousand words, a prototype is worth a thousand pictures". While Tom Kelley of IDEO in known to have said (Kelley 2001) "A picture is worth a thousand words. Only, at IDEO, we've found that a good prototype is worth a thousand pictures". And IDEO as an organisation frequently repeats the motto, "If a picture is worth a thousand words, then a prototype is worth a thousand meetings".

An example of a prototype in support of automotive design is presented in Figure 5.2. The complex and costly full-scale chassis prototype is typical of the artefacts which are used in testing and validation. The closer such prototypes are to the intended final artefact, the wider the range of issues which can be checked and the greater the number of tests which can be performed. While often expensive projects in their own right, such prototypes are deemed essential due

Figure 5.2 Example of an automotive prototype: full-scale chassis.
Source: Max Sims

to the need to detect and rectify errors as early as possible in the design process, avoiding costly later interventions. Such prototypes are routine elements of most design processes and can be determining factors in the success or failure of a project.

Science Fiction Prototypes

The use of physical and virtual prototypes is common in the final design and engineering stages of product development, but usually less so in the earlier stages. As Carleton and Cockayne (2009) have suggested, "While prototyping has a long history in the conceptualization and modelling stages of the innovation process, tangible prototypes that are intended to represent real opportunities have rarely existed in the fuzzy front end, much less at the vision stage". Early concept development has traditionally been the domain of verbal descriptions and of concept sketches.

But as the complexity of the products, systems and services has grown and their development costs have increased, the benefits of earlier prototyping have increased. Earlier, more detailed and more contextualised approaches are now being used to bring to light characteristics and implications which are not immediately obvious from the first verbal descriptions and concept sketches. New prototyping approaches are being developed as part of the effort to gain the maximum possible understanding, at the earliest point in time, as cheaply as possible.

One such approach is the science fiction prototype (Johnson 2010). Narratives suggest that it was first introduced at the Intel Corporation by the futurist Brian Johnson as part of the company's efforts to support product innovation. And that Johnson's short story *Nebulus Mechanisms* (Johnson 2009) of approximately 4000 words is the first true example of a science fiction prototype.

A science fiction prototype consists of a short vignette in text, theatre play, comic or video which describes an immersive future involving physical, psychological and sociological interactions (Johnson 2011). It typically adopts narrative structures and literary techniques as part of the effort to highlight the nature of the speculated future product, system or service. And unlike physical prototypes which are most often used to test form and function, science fiction prototypes help instead to test concepts and ideas.

In Johnson's words a science fiction prototype is " ... a short story, movie or comic based specifically on a science fact for the purpose of exploring the implications, effects and ramifications of that science or technology". He has suggested that the development and use of a science fiction prototype involve five main steps:

- Chose a Technology, Science or Issue: select a technology, science or issue which is to be explored by means of the science fiction prototype.
- Establish the World of the Story: establish the world of the story by introducing the reader or viewer to the people, locations and contexts.
- Introduce the Scientific Inflection Point: introduce the specific technology, science or issue which the reader or viewer is to explore.
- Explore the Ramifications: explore the implications and ramifications of the technology, science or issue on the world of the story (the people, locations and contexts).
- Ask What Did We Learn?: describe what was learned from placing the technology, science or issue into a realistic setting. Describe what could be modified to improve the impact on humans.

And Fergnani (2021) has suggested that science fiction prototypes always contain three key elements:

- treatment of the technology;
- treatment of the future individuals who are using/interacting with the technology;
- allusions to the future social context where the technology is embedded.

Most current science fiction prototypes are built using words from natural language, thus they exhibit all the strengths and weaknesses which characterise words. For example, words usually have more than one meaning. And even when a single meaning is evident, a word only rarely suggests more than a small number of attributes and characteristics. A word such as "automobile" may bring to mind four wheels, an engine of some type and a driver, but not much more. The form, colour, aesthetics, dynamics and usage cannot be directly determined from the single word.

Most words are semiotic symbols of relatively low descriptive power. Consideration of the many shapes and sizes of animal which can be alluded to with the word "dog" is sufficient to highlight the under-definition which can occur for artefacts, individuals or contexts. Such under-definition can lead to problematic misunderstandings during conversations, but in the case of science fiction prototypes can also lead to helpful divergent interpretations and to creative thinking due to the freedom which it allows.

But as Graham et al. (2013) have noted, science fiction prototypes are not straightforward to produce because "... they require the developer to be both a scientific or technological expert and also to be able to write compelling fictional stories". How to achieve narrative focus on the most important characteristics of the product, system or service is a subject of current debate. As is the choice of characters, lifestyles, emotions and other human considerations to highlight. And there is little or no standardisation at the moment in terms of the optimal writing style, length of narrative, depth of treatment or use of sector specific tropes.

One example of the challenges faced by the writers of science fiction prototypes is that of perspective. In their analysis of storyworlds, science fiction and future fictions, Von Stackelberg and McDowell (2015) suggested paying close attention to the four irreducible perspectives which people use to understand any aspect of the world:

- The subjective perspective examines the individual's interior world, with its concerns of perceptions, values, motivations, goals and the meaning of life.
- The objective perspective examines the individual's exterior world, with its concerns about changes in the ways people act externally.
- The interobjective perspective examines the collective exterior world, generally referred to as the physical world, with its concerns about measurable changes in natural and constructed external environments.
- The intersubjective perspective examines the collective interior world of the shared meaning of groups, as expressed in their culture, with concerns about shared collective structures, such as changes in languages, cultures, and institutions.

Ideally, a science fiction prototype should facilitate sense-making by supporting each of the four perspectives. Storylines which manage to support each would provide a greater width and depth of sense making. A rich and well-crafted science fiction prototype places the stakeholders in the imagined world and brings them face-to-face with the prospected new technology or new societal trend, stimulating comment, critique, suggestions or even redesign.

The reasons for the growing popularity of science fiction prototyping have been considered by several researchers. Michaud (2020) has suggested that:

Brainstorming is outdated, and design thinking has transformed into design fiction, i.e. the stimulation of the individual or collective imagination, with the aim of generating prototypes, in the form of objects or fictions, which will then be used to think about strategy and R&D activities in a better way.

And Underwood (2020) has claimed that:

Sci-fi prototyping has become big business. In recent years major multi-national companies like Nike, Google, Apple, Ford and Visa, and governmental bodies like NATO and the French army have all enlisted the services of sci-fi writers, commissioning conceptual futuristic narratives to help them imagine the worlds in which their products, services and strategies might very soon exist.

And one review performed by Kymäläinen (2016) identified the use of science fiction prototypes in areas including crowdsourcing, healthcare, 3D printing, cloud computing, augmented reality environments, brain-computer interfaces, biologically inspired computing, context-aware computing, artificial intelligence, sensor networks, internet of things, robotics and space exploration.

The use of science fiction prototypes for purposes of design is today becoming common. The convenience lies in providing a proving ground for testing concepts long before the designs can be manifested physically. As Graham (2013) has suggested:

… the outstanding distinction between story-led scenario planning and creative fiction prototypes is a distinction on which there seems to be general agreement. Scenario-based story telling is usually associated with a relatively short time span of up to five to ten years, while prototypes focus solely on the long-term time frames, typically ranging from ten to fifty years and sometimes even longer.

And, as the futurist sci-fi writer Madeline Ashby is reported (Underwood 2020) to have said about science fiction, "We can surface tensions and ideas and problems that maybe the institutional culture doesn't want to talk about".

Design Fictions

The first use of the term "design fiction" is usually credited to the science fiction author Bruce Sterling (2005) who later (2013) defined it as "Design fiction is the deliberate use of diegetic prototypes to suspend disbelief about change". In Sterling's use, the term "diegetic" refers to how objects or technologies can help support, or even generate, plot in literary works and films. Sterling sought to emphasise how objects create their own narratives in people's minds. On this view a design fiction is any object which can stimulate a story which can be communicated, debated and used as a basis for decision-making.

Lindley and Coulton (2015) instead defined design fiction more simply as something which:

- creates a storyworld;
- has something being prototyped within that storyworld;
- does so in order to create a discursive space.

The popularising of the term is often attributed to the work of Julian Bleecker who originally (2009) described it as:

Design fiction is a mix of science fact, design and science fiction. It is a kind of authoring practice that recombines the traditions of writing and story telling with the material crafting of objects. Through this combination, design fiction creates socialized objects that tell stories — things that participate in the creative process by encouraging the human imagination. The conclusion to the designed fiction are objects with stories. These are stories that speculate about new, different, distinctive social practices that assemble around and through these objects. Design fictions help tell stories that provoke and raise questions. Like props that help focus the imagination and speculate about possible near future worlds — whether profound change or simple, even mundane social practices.

Bleecker later (Bleecker et al. 2022) suggested that a work of design fiction is ideally the result of a process involving:

- collection of faint signals;
- selection of an archetype;
- presentation of stimulus materials;
- extrapolation from the signals;
- identifying the "what if";
- use of tropes;
- use of design workshops;
- making of the thing;
- dissemination;
- debate and reflection.

And Ahmadpour et al. (2019) have noted six typical characteristics which the resulting design fictions exhibit:

- They suspend disbelief in change.
- They help to reveal potential user concerns and uncertainties.
- They often examine the implications of a potential design or technology within the context of technological conflict.
- They enable discussion of the social and political context.
- They inspire discussion about desirable and preferable futures.
- They are not concerned with a finished design, being instead a disruptive space for emerging cultural artefacts.

Design fictions are thus physical artefacts which act as seeds for stories and for storytelling. Unlike prototypes which are used to evaluate form, function and user interactions, design fictions are instead artefacts which are deployed to stimulate thought and discourse. Viewing, touching, holding and using them stimulates reflections which support critical thinking, the spotting of challenges and the identifying of opportunities.

Developing and deploying a design fiction is thus an open activity which involves "what if" questions, physical artefacts and discourse. The physical artefacts help to examine the implications of a potential technology or of given societal trend, and help to reveal the associated concerns and uncertainties. Design fictions are thus enablers of discussions about preferable futures.

Kirby (2010) has observed that:

> Diegetic prototypes have a major rhetorical advantage even over true prototypes: in the fictional world – what film scholars refer to as the diegesis – these technologies exist as 'real' objects that function properly and which people actually use. Diegetic prototypes extend the analytical utility of virtual witnessing by addressing the issue of how cinematic depictions can lead to real-world technological development. The notion of virtual witnessing – that cinematic narratives present scientific and technological objects as conforming to natural reality – is central to a diegetic prototype.

The ability of design fictions to transcend the immediate and the instrumental has been considered in detail by Lindley (2018). Table 5.3 presents Lindley's comparisons between design fictions and the approaches which have traditionally been used in design ethnography.

TABLE 5.3 Comparing design ethnography to design fiction.

	Design Ethnography	**Design Fiction**
Temporal Dimension	the present	the future
Source of Context	situatedness	diagesis
Outputs	actionable insights	discursive space
Methods	established	under development

Source: Adapted from Lindley (2018).

As the comparisons suggest, while design ethnography captures mostly information about the current and the local, design fictions tend instead to stimulate discourse about the future, the relational and the unfolding. Lindley's analysis thus suggests the potential benefit of design fictions when selecting the characteristics of future artefacts which are likely to require new metaphors or new meanings.

Ideally, the artefacts which act as design fictions will transcend the immediate and the instrumental. They should not be part of the everyday world, but should instead be new artefacts which are deployed to focus, explore and question. As tangible material manifestations of something new and unusual, they will demand sense-making.

And among the many assumptions which may be challenged there will often be those of an ethical nature. Indeed, several researchers have discussed the interrelation between design fictions and ethics. Jensen and Vistisen (2017) have, for example, suggested that, "It can thus be argued that the narratives of design fictions should effectively make their ethical stances explicit and make them part of the discourse surrounding the storytelling around the diegetic prototype".

And Baumer et al. (2018) have suggested that:

> In some ways, all (design) fiction might be considered ethical, in so far as both ethics and fiction deal with what people, real or imagined, either should or would do in specific scenarios. For any issue, there are multiple different kinds of arguments that one could make about how to proceed ethically. Design fiction essentially allows us to take a designerly approach to issues of ethics and technology. In so far as design revolves around simultaneous consideration and exploration of different possible paths forward, design fiction provides an opportunity to consider different means of ethical judgments and decision making.

Over the years design fictions have been developed in many settings including the IKEA catalogue (Brown et al. 2016), voice assistants (Ringfort-Felner et al. 2022), digital personal assistants (Søndergaard and Hansen 2018), smart homes (Schulte et al. 2016), public libraries (Kozubaev and DiSalvo 2020), cities (Ylipulli et al. 2016), environmental interventions (Maxwell et al. 2019), wellbeing technology (Ahmadpour et al. 2019), dementia care (Noortman et al. 2019) and service design (Pasman 2016; Harwood et al. 2020) among others.

An example of a design fiction in support of automotive design is presented in Figure 5.3. The working artefact projects a holographic image of the driver outwards from the vehicle to a fixed distance in front. The virtual representative provides visual and gestural support to the driver's vocalisations during street interactions with other humans. The discourse resulting from the operation of the design fiction would likely explore the nature and ethics of such interactions, and would likely stimulate the ideation of additional possible human interaction approaches.

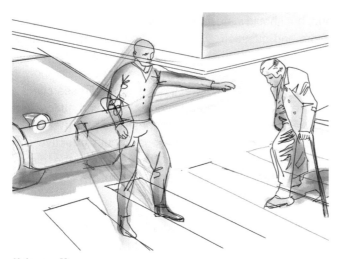

Hologram Me

Concept: holographic projectors at the front, rear and sides of a vehicle which can project an image of the vehicle driver to a distance of a few meters. The holographic images move and gesticulate appropriately to match the voicing which is emitted from the vehicle via loudspeakers.

Purpose: to interact with other road users, pedestrians, shop owners and others to communicate information to them in real time without the driver having to leave the vehicle.

Figure 5.3 Example of an automotive design fiction: Hologram Me.

Source: Max Sims

Speculative Design

The term "speculative design" is often said to have emerged in the 1990s at the Royal College of Art in London through the work of Anthony Dunne and Fiona Raby. It involves the use of artefacts which provoke speculations about possible futures, often with the intention of challenging the underlying assumptions of a technology or of a societal trend. It is one of several forms of "critical design", i.e. the practice of using design to critique technologies or societal trends which have been widely and uncritically accepted.

In *Speculative Everything* (2013) Dunne and Raby wrote that:

> Critical Design uses speculative design proposals to challenge narrow assumptions, preconceptions and givens about the role products play in everyday life. It is more of an attitude than anything else, a position rather than a method. There are many people doing this who have never heard of the term critical design and who have their own way of describing what they do. Naming it Critical Design is simply a useful way of making this activity more visible and subject to discussion and debate. Its opposite is affirmative design: design that reinforces the status quo.

And James Auger (2013) added that:

> Critical design uses designed artefacts as an embodied critique or commentary on consumer culture. Both the designed artefact (and subsequent use) and the process of designing such an artefact cause reflection on existing values, mores and practices in a culture.

Johannessen et al. (2019) have suggested that:

> Whereas traditional design often strives to make messages as clear as possible to enable users to think less, the speculative design practice attempts to do the opposite. In speculative design discursive practice, it is important to ask questions without dictating the audience's perception of an answer or a solution. Therefore, to enable the audience to entertain personal interpretations, speculative design scenarios are often open-ended, unclear, and complicated, and strive to provoke using dark humour and satire.

While Malpass (2017) has suggested that

> ... rather than presenting utopic or dystopian visions, speculative designs pose challenging statements that attempt to explore ethical

and societal implications on new science and the role product and industrial design plays in delivering this new science. The aim is to make scientific theories and the cultural implications of science perceptible in different ways.

Rather than being a commercial design practice, speculative design is instead a creative practice which aims to achieve awareness and discourse by means of artefacts which provoke. And rather than use functional prototypes, speculative design often instead uses para-functional proto-types (Dunne and Raby 2013) which are concrete materialisations which have unexpected and sometimes contradictory functions.

Para-functional prototypes do not necessarily simplify an action or a situation; instead they often complicate it. Ambiguity, counterfactuals, reductio ad absurdum and satire are used to cause feelings of confusion, unease and doubt, challenging people to rationally explain their innate or irrational responses. Through their inconsistencies and contradictions, para-functional prototypes aim to transgress. Their act of being in the world provokes and challenges.

Lukens and DiSalvo (2012) have suggested six properties which they claim are characteristic of speculative design projects:

- Speculative design projects are future oriented.
- Speculative design projects call attention to the possibilities and consequences of technological development and implementation.
- Speculative design projects involve and encourage thinking broadly about technology, beyond a single objective or practice.
- Speculative design projects are cross-disciplinary and integrative.
- Speculative design projects demonstrate an understanding of the interrelationship between information resources, technological structures, and market forces.
- Speculative design projects favour inquiry, experimentation and expression over usability and marketability.

And Johannessen et al. (2019) have suggested that speculative design can be used to explore issues via a three-step process:

- Define a topic or context for debate.
- Use "what if" questions to define alternative presents and futures which can be expressed as scenarios.
- Choose one or more of the scenarios and materialise them as either narratives, objects, or a combination of both, to provoke an audience.

Materialisation is in fact a key aspect of speculative design. At the heart of any speculative design activity lies some artefact of some form which provides tangible evidence of what is being discussed. The speculation is about some thing, not some idea. And it is the sensory, cognitive and emotional affordances of the thing which influence the discourse. The thing's act of being in the world is sufficient to provoke and challenge.

And a major aim of the thing is to achieve a degree of ambiguity. Ambiguity facilitates creativity. Ambiguity provides space for metaphors and meanings. As Gaver et al. (2003) have suggested,

> ... by thwarting easy interpretation, ambiguous situations require people to participate in making meaning. This can involve the integration of previously disconnected discourses, the projection of meaning onto an unspecified situation, or the resolution of an ethical dilemma. In each case, the artefact or situation sets the scene for meaning making, but doesn't prescribe the result. Instead, the work of making an ambiguous situation comprehensible belongs to the person, and this can be both inherently pleasurable and lead to a deep conceptual appropriation of the artefact.

Gaver et al. (2003) have described three forms of ambiguity which can play a role in design:

- Ambiguity of Information: ambiguity that arises due to the way that information is presented such as being incomplete or involving an unfamiliar language or number system.
- Ambiguity of Context: ambiguity that arises not because the thing itself is unclear, but because it can be understood differently in different contexts, with each context producing a different meaning.
- Ambiguity of Relationship: ambiguity that arises from imagining the interactions and possible relationship with the thing. Would it be used? Under what circumstances? What would life be like as a consequence?

And have stressed that:

> The most important benefit of ambiguity, however, is the ability it gives designers to suggest issues and perspectives for considera-tion without imposing solutions. Ambiguity of information impels people to question for themselves the truth of a situation ...

ambiguity is a powerful design tool for raising topics or asking questions, while renouncing the possibility of dictating answers.

But, of course, ambiguity has its limits. Exceeding the boundaries of physics or of human nature can hamper the sense-making and impede the construction of meaning. As Auger (2012) has warned,

> ... if a speculative design proposal strays too far into the future to present clearly implausible concepts or describes a completely alien technological habitat, the audience will fail to relate to the proposal, resulting in a lack of engagement or connection. In effect a design speculation requires a 'perceptual bridge' between the audience and the concept. Inspiration and influence can be drawn from diverse fields such as observational comedy, psychology, horror films and illusion, for the insights they offer into the complex workings of human perception and how it can be consciously manipulated to elicit reaction.

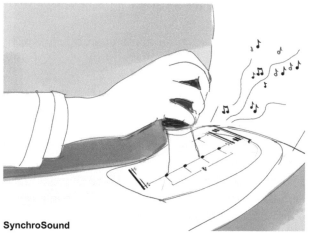

SynchroSound

Concept: a music moderation algorithm which oversees the playlist to match the beats-per-minute which the driver is performing when manually changing gears. Drivers take between 500 msand 1 second to perform a vertical change (1st–2nd, 3rd–4th, 5th–6th) and between 1 and 2 seconds to perform a horizontal change (2nd–3rd, 4th–5th). The selected songs are thus usually in the range from 30 bpm to 120 bpm.

Purpose: to better synchronise the environmental stimuli which the driver is perceiving. To increase the driving experience by bringing a further stimulus into the psychological driving flow.

Figure 5.4 Example of a para-functional prototype: SynchroSound.

Source: Max Sims

In summary, speculative design is more of an approach than a process or a methodology. A material artefact is usually involved, but its exact form is dependent on the topic which is being explored and on the designer who is doing the exploring. The intention is usually not to present a proposal but instead to critique a technology or societal trend whose characteristics and complexities have been widely and uncritically accepted. The intention is not to propose, but to provoke.

An example of automotive speculative design is presented in Figure 5.4. The para-functional prototype uses the gear shift to also select music of matching rhythm from the vehicle's infotainment system. Based not on functional driving needs but instead on the observation of the often excessive sophistication and cost of modern automotive systems, it highlights previously unnoticed similarities in the perceptual environment and unites them. The resulting user experience transcends normal functionality and normal expectation, provoking the driver to question the similarities and critique the vehicle's on-board technologies.

Speculating about Autonomous Road Vehicles

Forlano (2019) suggests that autonomous road vehicles are currently one of the many sociotechnical imaginaries which are present in our society. The term "sociotechnical imaginary" refers to a set of collectively held, institutionally stabilised and publicly performed visions of desirable futures involving advances in science, technology, social life and social order (see Jasanoff and Kim 2015 for details). Autonomous road vehicles have been sufficiently discussed and sufficiently developed by a sufficiently large number of people and organisations to have achieved socio-industrial momentum. Most existing robot-taxis and goods delivery robots have common characteristics which are based on specific assumptions about the power source, available road infrastructure, human need and robot role in society.

But Forlano has also noted the possibility of divergence between the current visions of corporate and governmental organisations on the one hand, and the eventual future outcomes on the other. In Forlano's words, "...speculative design interventions offer ways of resisting, disrupting, and destabilising the normative visions of linear technological progress toward an inevitable autonomous future". The autonomous road vehicle momentum can be directed, steered or even disrupted if the human needs can be accurately identified and clearly articulated.

If, as Forlano suggests, speculative design interventions can resist, disrupt and destabilise the normative visions of autonomous road vehicles, then the question becomes one of how best to do so.

A design space needs to be defined. A speculative design approach needs to be chosen. And speculations need to be produced.

Gero has suggested that there are routine designs which are situated within the well-trodden limits of the existing, designs which occupy a space of innovation, and a limiting outer barrier defined by constraints beyond which a design is not possible or not supportive of people (see Figure 5.5). The outer boundary is defined by the material, physical, psychological, sociological and ethical constraints of our real world. The outer boundary separates what is possible and useful from what is either not possible or not useful. Unlike the world of science fiction, the space for innovation in the real world is large, but not infinite.

The author's intention so far in this book has been to clarify as many as possible of the human-facing design considerations which affect the space of the innovative autonomous road vehicle designs. Each of the previous chapters has provided items of information which can help to locate the approximate boundaries of the innovation space.

Chapter 2 introduced the history of autonomous road vehicles to help establish the current region of routine autonomous road vehicle designs. Chapter 3 introduced instead a number of facts about the aesthetics, dynamics, behaviours, conversations, personalities and trust of autonomous road vehicles. Chapter 3 suggested likely human-facing limits of the region of innovative autonomous road vehicle designs. And Chapter 4 discussed a number of speculations which emerge from the world of science fiction. Chapter 4 provided several examples of frightening and ethically unacceptable autonomous road vehicles which must surely be beyond the limit of the possible designs.

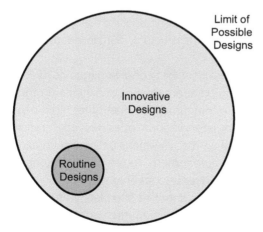

Figure 5.5 Routine designs and innovative designs.

Source: Adapted from Gero (1990)

The previous chapters have thus suggested rough and fuzzy boundaries between current routine designs, possible future innovative designs, and what is possible and desirable. They have attempted to describe as best possible the design space within which the designers will be working. And with a design space established, the priority passes to deciding how to explore that space.

Montgomery (2018) reviewed the most frequently cited speculative approaches and considered the logical interrelationships between them. A visual summary of the findings is presented as Figure 5.6. The diagram emphasises the wide-ranging nature of speculative design and its borrowing of approaches from other more specific domains such as from the worlds of science fiction, future studies, critical design and design fictions. Speculative design thus incorporates a set of individual approaches. And each of the individual approaches is best suited to a different time step into the future and to either choosing, understanding or critiquing.

This chapter has introduced five of the most popular and most commonly encountered future-facing design approaches. Scenarios, prototypes, science fiction prototypes, design fictions and speculative design have each been discussed and their main strengths and weaknesses noted. Each tends to be best suited to a different time step into the future. And there are obvious differences between them in terms of whether they were originally developed for purposes of

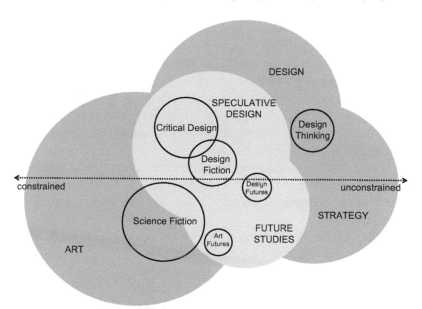

Figure 5.6 Speculative design in relation to other future-facing activities.

Source: Adapted from Montgomery (2018)

choosing, understanding or critiquing. Each is therefore very helpful in some applications but less so in others.

From among the future-facing design interventions discussed in this chapter the one which was judged by the author to be best suited to resisting, disrupting and destabilising normative visions of autonomous road vehicles was that of science fiction prototypes. The approach was judged to not suffer the material and manufacturing constraints of prototypes, design fictions and para-functional prototypes. And the approach was judged to be well suited to the 10 to 20 year time step into the future which had been chosen by the author for the speculations.

Specific considerations which led to the selection of science fiction prototypes as the primary future-facing design approach to adopt in the remaining chapters of this book include the following:

- The most revealing time step into the future is probably the point beyond which the friendly neighbourhood robots will be fully autonomous, i.e. from when they drive themselves with little or no human oversight, converse like humans and have human-like emotional capabilities. And from sector specialist opinions such sophistication seems likely to be approached in approximately ten to twenty years' time.
- Physical prototypes are usually developed for the present and scenario-based storytelling is usually associated with a relatively short time span of five to ten years into the future.
- Design fictions, para-functional prototypes and science fiction prototypes tend to be suited to time frames ranging from ten to fifty or more years into the future.
- Most forms of speculative design such as para-functional prototypes appear best suited to critiquing. They are ideal for noting consequences, particularly negative consequences. Less so for identifying desirable alternatives.
- Useful prototypes, para-functional prototypes and design fictions of friendly neighbourhood robots are likely to require some materials or some technologies which are not yet fully available at the current point in time.
- Useful science fiction prototypes can be inexpensively achieved by means of only natural language (words).

What now follows in this book are four chapters of science fiction prototypes of future friendly neighbourhood robots. The stories can help to contextualise their function and consider their operation. Hopefully the nature of the future friendly neighbourhood robots will emerge from the stories, as will many of the design constraints and design opportunities. The stories should help to support design discourse.

Each of the four chapters which follows begins with a summary of design-relevant observations, followed by four individual science fiction prototypes. The first of the chapters deals with robots which transport people, the second with robots which provide workplaces, the third with robots which provide healthcare and the fourth with robots which provide entertainment. Each chapter provides a relatively wide introduction to the opportunities and challenges in relation to the service provision which is being considered.

When considering 20th-century architecture Ligo (1973) identified five "functions" which he claimed were repeated topics of 20th-century discourse. Buildings and other major architectural works were repeatedly subjected to five types of scrutiny:

- Structural Articulation: a building's materials, construction techniques, human activities and layout of spaces.
- Physical Function: the utilitarian tasks which can be performed in the building and the building's ability to achieve the wellbeing of its occupants.
- Psychological Function: the emotional responses and subjective feelings elicited by the building, and its wider effects on the human psyche.
- Social Function: the capacity of the building to support social interactions and to provide a meaningful milieu for people.
- Cultural-Existential Function: the capacity of the building to concretise humanity's universal values, transcending an individual's life at a specific time and specific place.

While perhaps not strictly correct to speak of psychological properties, sociological properties or value systems as "functions", the pragmatic approach nevertheless has a logic to it and helps shed light on the human interactions involved. The remaining chapters of this book will hopefully shed some light, explicitly or implicitly, on a few of the "functions" of the future friendly neighbourhood robots.

Having set the scene, it is now time to speculate.

References

Ahmadpour, N., Pedell, S., Mayasari, A. and Beh, J. 2019, Co-creating and assessing future wellbeing technology using design fiction, *She Ji: The Journal of Design, Economics and Innovation*, Vol. 5, no. 3, pp. 209–230.

Alcamo, J. 2007, Methods for building scenarios of the environment, Lecture by Joseph Alcamo, Peyresq, France, 1–13 September.

Alcamo, J. and Ribeiro, T. 2001, *Scenarios as tools for international environmental assessment*, Vol. 5., European Environment Agency.

Auger, J.H. 2012, *Why Robot? Speculative design, the domestication of technology and the considered future*, PhD thesis, Royal College of Art, London, UK.

Auger, J. 2013, Speculative design: crafting the speculation, *Digital Creativity*, Vol. 24, No. 1, pp. 11–35.

Baumer, E.P., Berrill, T., Botwinick, S.C., Gonzales, J.L., Ho, K., Kundrik, A., Kwon, L., LaRowe, T., Nguyen, C.P., Ramirez, F. and Schaedler, P. 2018, What would you do? Design fiction and ethics, Proceedings of the 2018 ACM Conference on Supporting Groupwork, Sanibel Island, Florida, USA, 7–10 January, pp. 244–256.

Bleecker, J. 2009, *Design Fiction: a short essay on design, science, fact and fiction*, Near Future Laboratory, USA.

Bleecker, J., Foster, N., Girardin, F. and Nova, N. 2022, *The Manual of Design Fiction*, Near Future Laboratory, Venice, California, USA.

Brown, B., Bleecker, J., D'adamo, M., Ferreira, P., Formo, J., Glöss, M., Holm, M., Höök, K., Johnson, E.C.B., Kaburuan, E. and Karlsson, A. 2016, The IKEA Catalogue: design fiction in academic and industrial collaborations, Proceedings of the 19th ACM International Conference on Supporting Group Work, November, pp. 335–344.

Carleton, T. and Cockayne, W. 2009, The power of prototypes in foresight engineering, Proceedings of the 17th International Conference on Engineering Design ICED 09, Vol. 6, Part 2, Design Methods and Tools, Palo Alto, California, USA, 24–27 August.

Carroll, J.M. 1997, Human–computer interaction: psychology as a science of design, *International Journal of Human-Computer Studies*, Vol. 46, No. 4, pp. 501–522.

Carroll, J. 2000, Five reasons for scenario-based design, *Interacting with Computers*, Vol. 13, No. 1, pp. 43–60.

Cha, K. 2019, *Affective Scenarios in Automotive Design: a human-centred approach towards understanding of emotional experience*, doctoral dissertation, Brunel University London, London, UK.

Cooper, A. 1998, *The Inmates Are Running the Asylum: why high tech products drive us crazy and how to restore the sanity*, SAMS, Pearson Education, Indianapolis, Indiana, USA.

Dunne, A. 2008, *Hertzian Tales: electronic products, aesthetic experience, and critical design*, MIT Press, Cambridge, Massachusetts, USA.

Dunne, A. and Raby, F. 2013, *Speculative Everything: design, fiction, and social dreaming*, MIT Press, Cambridge, Massachusetts, USA.

Fergnani, A. 2021, Science fiction prototyping, design fiction and worldbuilding, what differences?, *Medium*, https://medium.com/predict/sci-fi-prototyping-design-fiction-and-worldbuilding-what-differences-ca6a5bbd189.

Floyd, C. 1984, A systematic look at prototyping, in R. Budde, K. Kuhlenkamp, L. Mathiassen and H. Züllighoven (eds) *Approaches to Prototyping*, Proceedings of the Working Conference on Prototyping, 25–28 October 1983, Namur, Belgium, Springer-Verlag, Berlin, pp. 1–19.

Forlano, L. 2019, Stabilizing/destabilizing the driverless city: speculative futures and autonomous vehicles, *International Journal of Communication*, Vol. 13, pp. 2811–2838.

Gaver, W.W., Beaver, J. and Benford, S. 2003, Ambiguity as a resource for design, in Proceedings of the SIGCHI Conference on Human Factors in Computing Systems (CHI'03), Fort Lauderdale, Florida, USA, 5–10 April, pp. 233–240.

Gero, J.S. 1990, Design prototypes: a knowledge representation schema for design, *AI Magazine*, Vol. 11, no. 4, pp. 26–26.

Giacomin, J. 2017, What is design for meaning?, *Journal of Design, Business & Society*, Vol. 3, No. 2, pp. 167–190.

Giacomin, J. 2022, *Humans and Autonomous Vehicles*, Routledge, Abingdon, Oxon, UK.

Graham, G. 2013, Exploring imaginative futures writing through the fictional prototype 'crime-sourcing', *Futures*, Vol. 50, pp. 94–100.

Graham, G., Greenhill, A. and Callaghan, V. 2013, Exploring business visions using creative fictional prototypes, *Futures*, Vol. 50, pp. 1–4.

Harwood, T., Garry, T. and Belk, R. 2020, Design fiction diegetic prototyping: a research framework for visualizing service innovations, *Journal of Services Marketing*, Vol. 34, No. 1, pp. 59–73.

Houde, S. and Hill, C. 1997, What do prototypes prototype?, in M.G. Helander, T.K. Landauer and P.V. Prabhu (eds), *Handbook of Human–Computer Interaction*, North Holland, Elsevier, Amsterdam, Netherlands, pp. 367–381.

Jasanoff, S. and Kim, S.H. 2015, *Dreamscapes of Modernity: sociotechnical imaginaries and the fabrication of power*, University of Chicago Press, Chicago, Illinois, USA.

Jensen, T. and Vistisen, P. 2017, Ethical design fiction: between storytelling and world building, *The ORBIT Journal*, Vol, 1, No. 2, pp. 1–14.

Johannessen, L.K., Keitsch, M.M., Pettersen, I.N. 2019, Speculative and Critical Design – features, methods, and practices, Proceedings of the Design Society 22nd International Conference on Engineering Design (ICED19), Delft, Netherlands, 5–8 August.

Johnson, B.D. 2009, Nebulous mechanisms, in V. Callaghan, A. Kameas, A. Reyes, D. Royo and M. Weber (eds), *Intelligent Environments 2009*, Proceedings of the 5th International Conference on Intelligent Environments, Barcelona, Spain, 20–21 July, IOS Press, pp. 9–18.

Johnson, B.D. 2010, *Screen Future: the future of entertainment, computing and the devices we love*, Intel Press, Hillsboro, Oregon, USA.

Johnson, B.D. 2011, *Science Fiction Prototyping: designing the future with science fiction*, Synthesis Lectures on Computer Science 3, Morgan & Claypool Publishers, San Rafael, California, USA.

Kahn, H. and Wiener, A.J. 1967, The Next Thirty-Three Years: a framework for speculation, *Daedalus*, Vol. 96, No. 3, pp. 705–732.

Kelley, T. 2001, Prototyping is the shorthand of innovation, *Design Management Journal*, Vol. 12, no. 3, pp. 35–42.

Kirby, D. 2010, The future is now: diegetic prototypes and the role of popular films in generating real-world technological development, *Social Studies of Science*, Vol. 40, No. 1, pp. 41–70.

Kozubaev, S. and DiSalvo, C. 2020, The Future of Public Libraries as Convivial Spaces: a design fiction, Proceedings of the 2020 ACM International Conference on Supporting Group Work, Sanibel Island, Florida, USA, 6–8 January, pp. 83–90.

Kymäläinen, T. 2016, Science fiction prototypes as a method for discussing socio-technical issues within emerging technology research and foresight, *Athens Journal of Technology & Engineering*, Vol. 3, No. 4, pp. 333–347.

Ligo, L.L. 1973, *The Concept of Function in Twentieth-Century Architectural Criticism*, PhD thesis, University of North Carolina, Chapel Hill, North Carolina, USA.

Lim, Y.K., Stolterman, E. and Tenenberg, J. 2008, The Anatomy of Prototypes: prototypes as filters, prototypes as manifestations of design ideas, *ACM Transactions on Computer–Human Interaction (TOCHI)*, Vol. 15, No. 2, pp. 1–27.

Lindley, J. 2018, *A Thesis About Design Fiction*, PhD thesis, Lancaster University, UK.

Lindley, J. and Coulton, P. 2015, Back to the Future: 10 years of design fiction, in ACM Proceedings of the 2015 British HCI Conference, Lincoln, Lincolnshire, UK, 13–17 July, pp. 210–211.

Lukens, J. and DiSalvo, C. 2012, Speculative design and technological fluency, *International Journal of Learning and Media*, Vol. 3, No. 4, pp. 23–40.

Malpass, M. 2017, *Critical Design in Context: history, theory, and practice*, Bloomsbury Visual Arts, London, UK.

Manzini, E. 2003, Scenarios of sustainable well-being, *Design Philosophy Papers*, Vol. 1, No. 1, pp. 5–21.

Manzini, E. 2015, *Design, When Everybody Designs: an introduction to design for social innovation*, MIT Press, Cambridge, Massachusetts, USA.

Maxwell, D., Pillatt, T., Edwards, L. and Newman, R. 2019, Applying design fiction in primary schools to explore environmental challenges, *The Design Journal*, Vol. 22, No. 1, pp. 1481–1497.

Michaud, T. 2020, Design fiction, technotypes and innovation, in T. Michaud (ed.) *Science Fiction and Innovation Design*, John Wiley & Sons, Hoboken, New Jersey, USA, pp. 113–162.

Mitrović, I., Auger, J., Hanna, J. and Helgason, I. (eds) 2021, *Beyond Speculative Design: past–present–future*, SpeculativeEdu, Arts Academy, University of Split, Zagreb, Croatia.

Moggridge, B. 2007, *Designing Interactions*, MIT Press, Cambridge, Massachusetts, USA.

Montgomery, E.P. 2018, An unresolved mapping of speculative design, https://www.epmid.com/projects/Mapping-Speculative-Design.

Mulder, S. and Yaar, Z. (2006), *The User Is Always Right: a practical guide to creating and using personas for the web*, New Riders Publishers, Berkeley, California, USA.

Noortman, R., Schulte, B.F., Marshall, P., Bakker, S. and Cox, A.L. 2019, HawkEye – Deploying a Design Fiction Probe, Proceedings of the 2019 CHI Conference on Human Factors in Computing Systems, Glasgow, Scotland, UK, 4–9 May, pp. 1–14.

Ogilvy, J. and Schwartz, P. 1996, Plotting your scenarios, in J. Ogilvy (ed.), *Facing the Fold: essays on scenario planning*, Triarchy Press, Axminster, Devon, UK, pp. 11–34.

Pasman, G. 2016, Design fiction as a service design approach, in Service Design Geographies, Proceedings of the ServDes. 2016 Conference, May, No. 125, Linköping University Electronic Press, pp. 511–515.

Pruitt, J. and Grudin, J. 2003, Personas: practice and theory, in ACM Proceeding of the 2003 Conference on Designing for User Experiences DUX03, San Francisco, California, USA, 6–7 June, pp. 1–15.

Ringfort-Felner, R., Laschke, M., Sadeghian, S. and Hassenzahl, M. 2022, Kiro: a design fiction to explore social conversation with voice assistants, *Proceedings of the ACM Journal on Human–Computer Interaction*, Vol. 6, No. 33, pp. 1–21.

Rosson, M. and Carroll, J. 2002, *Usability Engineering*, Morgan Kaufman Publishers, San Francisco, California, USA.

Rudd, J., Stern, K. and Isensee, S. 1996, Low vs. high-fidelity prototyping debate, *Interactions*, Vol. 3, No.1, January, pp. 76–85.

Schulte, B.F., Marshall, P. and Cox, A.L. 2016, Homes for Life: a design fiction probe, Proceedings of the 9th Nordic Conference on Human–Computer Interaction, 23–27 October, Gothenburg, Sweden, pp. 1–10.

Schwartz, P. 1996, *The Art of the Long View: planning for the future in an uncertain world*, Bantam Doubleday, New York, New York, USA.

Sefelin, R., Tscheligi, M. and Giller, V. 2003, Paper Prototyping – What Is It Good For? a comparison of paper and computer based low-fidelity prototyping, In CHI'03 Extended Abstracts on Human Factors in Computing Systems, Fort Lauderdale, Florida, USA, 5–10 April, pp. 778–779.

Søndergaard, M.L.J. and Hansen, L.K. 2018, Intimate Futures: staying with the trouble of digital personal assistants through design fiction, ACM Proceedings of the 2018 Designing Interactive Systems Conference, Hong Kong, China, 9–13 June, pp. 869–880.

Sterling, B. 2013, Patently untrue: fleshy defibrillators and synchronised baseball are changing the future, *Wired*, October, https://www.wired.co.uk/article/patently-untrue.

Suchman, L., Trigg, R. and Blomberg, J. 2002, Working Artefacts: ethnomethods of the prototype, *The British Journal of Sociology*, Vol. 53, No. 2, pp. 163–179.

Underwood, K. 2020, The future is now, *Pivot Magazine*, Chartered Professional Accountants Canada (CPA), March/April, pp. 5052.

Van Notten, P. 2005, Scenario Development: a typology of approaches, Chapter 4 of OECD Report on Think Scenarios, Rethink Education.

Von Stackelberg, P. and McDowell, A. 2015, What in the world? Storyworlds, science fiction, and futures studies, *Journal of Futures Studies*, Vol. 20, No. 2, pp. 25–46.

Wilkinson, A. and Kupers, R. 2014, *The Essence of Scenario*, Amsterdam University Press, Amsterdam, Netherlands.

Ylipulli, J., Kangasvuo, J., Alatalo, T. and Ojala, T. 2016, Chasing Digital Shadows: exploring future hybrid cities through anthropological design fiction, Proceedings of the 9th Nordic Conference on Human–Computer Interaction, 23–27 October, Gothenburg, Sweden, pp. 1–10.

Chapter 6

Providing Transport

Background

For most of those who work in the field of autonomous road vehicles the current objective is to evolve the existing human-driven automobiles, taxis, passenger vans or buses into self-driving equivalents. Replacing the human driver with an automation of some form is the goal. And transporting people is arguably the main focus.

Adding the self-driving capability while largely maintaining the existing metaphor of the human-driven vehicle provides an important simplification. One which allows the engineers to focus on the automated driving rather than the many other aspects and opportunities of the vehicle. Changes can be limited to the vehicle's sensing, detection and control systems or to road features such as the lane markings, route selection or infrastructural adaptations.

In most cases the autonomous road vehicles which are currently being trialled are not distant from the existing human-driven taxis, airport shuttles, buses or trams in terms of size, aesthetics, function, ritual and meaning. Leaving the vehicle packaging and service considerations for a later date has to some extent been the norm.

While challenging technically, the handing over of the driver's responsibilities to the automation is straightforward psychologically. Such a transition does not usually prove particularly confusing to customers and is not likely to significantly disrupt the taxi, bus or other traditional metaphor. With the possible exception of an initial unease caused by the lack of a human driver, the travel experience of the autonomous pods and shuttles which are currently being trialled has proved to be roughly similar to that of human-driven equivalents. Often described as "normal" (see, for example McArdle 2022) and sometimes even as "boring" (see, for example, Standage 2018), the experience of being transported by current autonomous road vehicles is considered unexceptional by many. A greater incidence of motion sickness (see

DOI: 10.4324/9781032724232-6

Diels and Bos 2016 and Iskander et al. 2019 for reviews), but otherwise unexceptional.

There has been an air of "incremental" to the experiments in transporting people via autonomous road vehicles. The unique selling proposition (USP) has usually been prospected to be increased road safety and a reduction in running cost. Improving safety while reducing the cost and plugging gaps in existing travel networks would appear to be the current promise. A few recent concept cars have gone slightly further by taking full advantage of the available space and new design freedoms to introduce additional levels of comfort and luxury. For example, Renault's EZ-Ultimo Car Concept (Marchese 2018) explores several opportunities for more personalised and luxurious travel. But, overall, most current concepts involve a reworking of the existing human-driven vehicle metaphor rather than an outright replacement in terms of characteristics and expectations.

Such low-risk early offerings are likely to be among the first friendly neighbourhood robots to reach production and to begin supporting humanity. In fact, services such as traditional taxi fleets have been among the early adopters of friendly neighbourhood robots (see, for example, Kipp et al. 2020; Kim et al. 2020; Wikipedia Contributors 2023b). As have the operators of shuttles and busses (see, for example, Abe 2019; Bucchiarone et al. 2020; Lutin 2018).

But while most current autonomous taxis, airport shuttles, buses or trams are not significantly different from traditional human-driven road vehicles, more substantial deviations are on the horizon. Once the core capability of self-driving becomes safe and reliable it is inevitable that substantially different road vehicle metaphors will be experimented. A well-understood and possibly fully standardised autonomous road vehicle chassis will provide a solid platform for experimenting variations in the transporting capabilities and in the service provision.

For example, one frequently discussed concept is that of the "hotel robot" or "room on wheels" which provides hotel-like conveniences to travellers who are either connecting between different modes of transport or who are engaged in long-distance road travel. Such friendly neighbourhood robots bridge the current gap between the mobility aspect of the travel and the sojourn aspect. The proposals aim to simplify the travel experience by combining a part of travel and a part of rest, reducing the number of check-ins, saving transfer time and providing enhanced personalisation.

Typical of such friendly neighbourhood robots is the "Autonomous Travel Suite" conceived by Aprilli Design Studio (Turner 2022). The "Autonomous Travel Suite" is a concept for a mobile hotel room which features a bed, seating area, minibar, kitchen, bathroom, and storage for clothes and luggage. It is proposed in several sizes and room types which would have different costs. Customised equipment such as TVs or extra beds are also available upon request.

Beyond the simplifications and efficiencies offered to the customer, such proposals also stimulate interesting questions for urban planners. For example, are mobile rooms potentially less demanding on prime city-centre acreage and less expensive than more permanent and more rigid brick-and-mortar structures? And might such rooms on wheels prove easier to integrate into train, ferry or other forms of transport than previous offerings? Might "hotel robots" provide an additional opportunity for combining flexibility, privacy and possibly also luxury?

Another frequently discussed concept is that of "inclusive mobility robot". An inclusive mobility robot is a road vehicle which is specifically optimised to accommodate the widest possible range of physical, perceptual and cognitive characteristics and the greatest number of human mobility needs. Such a purpose and meaning is not entirely new with road vehicles, but takes on greater significance with friendly neighbourhood robots.

Language is pregnant with phrases which refer to friction, barriers and exclusion. Metaphors such as "being born on the wrong side of the tracks" attempt to convey the experiences of people who encounter impediments to their rights or ambitions. And terms such as oligarchism, elitism, racism, sexism, colourism and ableism illustrate the variety of characteristics and conditions which can be used to erect obstacles.

Autonomous road vehicles which are designed specifically to optimise accessibility, inclusivity and human dignity thus seem a reasonably certain prospect given their likely flexibility and probable cost effectiveness. Such services have been suggested to be potentially transformative opportunities (see, for example, Riggs and Pande 2021; Nanchen et al. 2022; Yousfi and Métayer 2022). And several early experimental designs of inclusive autonomous road vehicles are already being tested. An interesting example of the requirements which emerge when designing with accessibility, inclusivity and human dignity in mind is the case of Toyota's 2020 E-Palette Olympic Games shuttle (Hitti 2019). Ethnographic interactions with the athletes identified the need for larger doors and lower floors than those of a standard shuttle, and the need for electric ramps to facilitate ingress and egress for people, particularly those using wheelchairs.

And whereas physical, perceptual and cognitive barriers are being removed by current inclusive mobility robot designs the longer-term future also brings opportunities for addressing emotional, psychological and social barriers. Future friendly neighbourhood robots which are conversationally and emotionally adept will prove ideal platforms for services which provide psychosocial support (see Lim et al. 2021 for details). Speculations about social care robots which ferry vulnerable individuals to medical centres or social gatherings are one example of such reasoning. And speculations regarding relaxation robots and similar commercial services also belong to this particular form of friendly neighbourhood robot.

Automation may even lead to road vehicles for which the mobility function is secondary. Following in the footsteps of the human-driven bookmobiles and mobile libraries (Ortwein 2015; Wikipedia Contributors 2023a) of the past, some friendly neighbourhood robots may end up acting more as a venue than as a form of mobility. As a gathering place for psychological support, some friendly neighbourhood robots may replace brick-and-mortar venues for lifestyle, social or even religious activities.

Autonomous taxis, autonomous airport shuttles, autonomous buses, autonomous trams, hotel robots, rooms on wheels, inclusive mobility robots, relaxation robots, psychological support robots and social care robots are all forms of friendly neighbourhood robot which seem somewhat likely to appear on roads in the coming years. Such developments are likely consequences of the increasing demand for cost effective, flexible and personalisable mobility.

What follows on the remaining pages of this chapter are a set of four science fiction prototypes, short immersive vignettes which explore the nature and implications of the future friendly neighbourhood robots. Each involves physical, psychological and social interactions with the future friendly neighbourhood robot and several allude to anticipated ethical conundrums.

As discussed earlier in this book the science fiction prototypes are linguistically based and avoid an excessive focus on the robot itself, so as to leave ample room for imagination and speculation. It is the author's hope that they can provide a focal point for discussion of autonomous road vehicle design.

Ace Taxis

The barrister rose from her seat and approached the judge's bench, ready to articulate the perniciousness of the thoughtless taxi. And with a solemn but determined demeaner she began.

"I would like, my lordship, to now read the plaintiff's statement as she cannot be present in this courtroom today for the reasons previously expressed.

"As the only witness to the full set of events of the night of 21 December the plaintiff's testimony is central to the exposition and sheds significant light on the errors and lost opportunities which characterise the negligence which this case exposes.

"In deference to this truth I will read the statement in its entirety and will remain faithful to the text as I find it here before me.

"I reserve my annotations and precisions, my lordship, as legal counsel to the plaintiff, to my later clarifying statements.

"Upon your permission I will proceed to read the plaintiff's statement in its entirety".

I, Eva Safer, here state what happened on the night of 21 December when I had planned to meet friends at the Red Lion pub near Hathersage. I used my entangler to order the first taxi on the list, an Ace Taxi, which arrived about 15 minutes later. At first everything seemed ok, and the taxi took me out of town and into the Peak District as expected. Nothing much happened for about twenty minutes as I talked with mates of mine using my entangler to pass the time.

The taxi then drove up a narrow road with no lighting which I did not recognise. I thought to myself, odd this, since I know these parts well but hadn't been this way before. Soon the taxi began to say that we would be arriving in five minutes. Then it came to a full stop near a large barn, and opened its doors.

I tried to tell the thing that a barn is not a pub, and that we were nowhere near where I was supposed to be, but it simply repeated something about having checked its map coordinates and being at the right address. And I remember that it kept insisting that I get out, saying something about the ride being over and it having another call.

And then I made the mistake of getting out. I could not see anything from inside so I got out to have a look around in the darkness in case there was something which would convince the thing that we were in the wrong place. Then, quickly I think, it closed its doors and drove off despite my yelling my head off telling it to stop. It simply drove off. And left me there at nearly 9.30 on a cold and humid night. And I was soon shivering.

I was upset and used my entangler to contact the company to tell them to get another taxi over to pick me up before I froze in my evening dress and smart shoes. A narrow road in the middle of nowhere was no place for someone like me on a night like that. But I was not able to get through.

So I began walking back down the road the way the taxi had come. I walked for a long time. I don't know how long. And kept trying to get through to the company on the entangler. I couldn't see a house or shop or pub or person anywhere, and couldn't raise anybody.

Finally, I got through to the taxi people and one of them spoke to me. It was a man who called himself Ian. But now, after bringing this matter to court, the company is saying that it has no employee named Ian. Anyway, I remember clearly that he said his name was Ian, and that he would check the taxi's whereabouts and send it back to pick me up again. But no taxi came.

And after another half hour or more of walking in the cold I tried entangling again, but this time the company's virtual just said something about experiencing a busy period and about getting back to me as soon as possible. But they never got back to me.

By that point I was wet and shivering and miserable and had to walk down to the nearest people, which turned out to be at the Millstone pub. I later checked the distance and it was three miles. Three miles on foot at night in the cold. From the Millstone I entangled with another taxi company and they picked me up and took me home.

On 22 December the infection started (see attached medical note signed by Doctor Jenkins) and added to my existing condition (see letter of 10 May from Doctor Jenkins) and things got worse until I found myself in the situation which I am in today (see Doctor Sekaran's medical evaluation of 1 June).

All this should not have happened and should not be allowed to happen to others.

Ace Taxis failed me by sending me a robo-taxi which had some type of problem. I do not believe the claim that the drop-off error was in the digital map which the robot reads. The owners of the Millstone pub say that other taxi companies drop people off at their premises all the time but that nothing similar had ever happened. And I have recently learned that other taxi companies use the same map (see the attached letters from Alpha Taxies and Tiger Taxis).

Ace Taxis failed me by using a robo-taxi which first misread its map then did not correct itself once I told the thing that it was in the wrong place. And when I yelled at it to stop it either could not hear me or instead decided to not stop. It should have come back for me.

And Ace Taxis failed me in not providing a way to get through to them to sort the mess out once the robot had gotten it all wrong. Not having an emergency entangler channel is inexcusable, in my opinion. And dangerous.

The failures of Ace Taxis have caused the health conditions which are described in Dr Sekaran's evaluation. I thus feel that the company should take responsibility and remunerate me for my lost income and my ongoing medical expenses. And I feel that the company's operating licence should be reviewed in light of the dangers to the travelling public.

Eva Safer.

AllRide

Washington D.C. was made for processions and parades and grew with the automobile but that is no guarantee of success. Thankfully though on this occasion there were no double-parked cars, no broken-down vehicles and no drunks crossing the road. It was late evening, the traffic was behaving and there was noticeable relief among the remote operators as it glided down the wide avenue and made its way to destination.

Right on time AllRide showed up at the correct address just off Wisconsin Avenue and found an open kerb. This is of course not always the case. There is nothing which annoys passengers and remote operators more than a shuttle stopping too far from the front door. But this evening the kerb gods smiled down upon the mortals and their friendly neighbourhood robot. The sidewalk seemed free enough. And it seemed smooth enough. And there was no clumsy sticky gate lock to negotiate. Or steps. The dreaded steps. If only all missions were so straightforward.

Annette emerged from the doorway with a slow and calculated movement. She was sitting on a particularly sophisticated device, maybe best called a wheelchair. It was equipped with a throng of things which the orthopaedists considered useful, and must have been a rather expensive piece of kit. Probably a godsend for its owner. But maybe not for the transport shuttles. This one appeared wide, unusually wide, with tyres and fender guards and sensors sticking out. The sight of it triggered groaning and tutting among the shuttle operators in their cramped little office.

Annette knew the routine. She had used many shuttles over the years to travel to work, to enjoy evenings out and to rush to the side of sick friends. She had known from an early age that her legs would not prove very helpful, and had come to accept that wheelchairs and shuttles might be constant companions. Years of practice had honed her skills and perfected the drill.

This specific evening she had cut her timings a bit short. A considerable risk given that missing the meet-up would lead to a litany of complaints. God knows how Carmen could be a pain. How pedantic she seemed to be about meeting times. A minute or two of delay would alchemise into an hour or two of whining. Thus on this occasion an AllRide was required.

A safety measure of sorts. No delays, no mistakes, no messing around. Best in town.

The shuttle arrived at the declared time, exactly, to the second. It slowed to a stop and began to quietly open side doors. For people who regularly used wheelchairs there was something reassuring about those large red doors. A welcoming smile of sorts. A big grin, a promise of ease of entry, lack of fuss and normalcy. Even the soft ringing sounds which accompanied the opening door seemed to whisper, "You will get to the Blue Moon on time". The situation was in hand. Everything would be fine. Normalcy.

Then the bricks appeared in the corner of Annette's eye. Blood red, hard and spikey. The mound partially obstructed the garden walk as it perched half on the walk and half on the rose bed. The sight washed across Annette's previously placid mind like an incoming tide. Forgot to tell the workmen. Oops. Can I pass? Not good. Should be out back. Crap. Not good. Can I pass? The same mound provoked an audible sound which leaked from the remote operators as their screens revealed the obstacle. And as AllRide's rooftop camera brought the challenge fully into view, the leak burst into a full-scale flow. "This may take a while", moaned one of the operators.

The Little Helper was a work of art in its own right. A small tracked robot with helpful arms and handles and gripping points. A tour de force of ergonomic and orthotic design. Little Helper was able to get nearly anybody out of a jam. At least on a sunny day. When it rained or snowed things could sometimes get tricky, but in good weather the Little Helper always got the job done. And, fortunately, it did not snow all that often in Washington D.C.

The Little Helper usually remained in its cubby hole at the back when not needed. Small, hidden and unnoticed. A capability ready to be capable. Little Helper was the synthetic superhero who came to the rescue when irregular sidewalks, obstacles or other inconveniences were detected. When AllRide spotted trouble, the Little Helper was deployed without any intervention from the humans. If shuttles think, then AllRide took great pleasure in showing the humans who was boss.

This time, however, AllRide could not detect the difficulties with its usual sensors due to the bricks being out of its line of sight. AllRide was lowriding. Sitting too low on the road to see everything it needed to see. The human operators thus had to spot the problem from their rooftop backup camera.

"The lousy shuttle never sees anything important", said one of them.

"Neither do you", said the other, before hitting the button to launch Little Helper on its way.

"Little Helper deployed", confirmed the remote monitoring screen.

And "Little Helper on its way please wait for assistance", urged the big screen on AllRide's roof.

Once the Little Helper had clinked and clanked past the florals and come to a stop, Annette fumbled for grip, leaning this way then that. Hands reached, fingers grabbed, weight was transferred and a tow ensued. Not an elegant way to start a night on out on the town, she thought, but the clock was ticking and there was little time for niceties. The scene, reminiscent of a rescue at sea, concluded at AllRide's door after what seemed like half a day but must instead have been only half a minute.

Once through AllRide's doors the usual grips, handrails and cords deployed and provided all the hold that Annette could ask for. Since she had no guide dog, comfort pet or other biological buddy with her, AllRide folded back its animal panels as soon as its sensors gave the all-clear. Annette then moved into the recommended position and fastened her wheelchair to the safety rail.

"We're in", she said out loud, as if expecting AllRide to reply.

But the shuttle could not reply. AllRide shuttles buzz, chime, ding, ring and shriek, but don't talk. Company rules forbade talking. Too likely to be misinterpreted, too likely to be misunderstood, too likely to lead to insult. Too risky.

Annette waved into the main passenger camera imagining that someone, somewhere, might be waving back in reply. One of the outdated rituals which humans were stubbornly clinging to despite its obvious obsolesce in the autonomous era. It's a funny feeling, she thought, to never know who to thank.

"She's off with plenty of time to spare", said the first remote operator. "If she had got stuck in that garden who knows how long it would have taken us to free her".

"Yeah", replied the second. "And if we had taken longer than regulation you know who the jerks at head office would have blamed for the screw-up".

"Of course", said the first. "The sods always side with the shuttle".

Social Sara

Hiroshi pulled his jacket tight and fought the aqueous assault while pacing and fidgeting at the street corner. It was a damp evening in North London and the wet and chill were now competing with the traffic for his attention. Filthy weather he thought. It better arrive soon or they had better bring a change of clothes.

He had deliberated the decision for weeks and the matter had become tiring. At times it seemed a great idea for meeting people in his newly adopted city. Just give it a go. Just meet a few people. At other times it seemed like a recipe for embarrassment. After all, what was a young man from southern Japan supposed to say to a Londoner? Do you like sushi? Wearily, and with enough doubts to fill a shipping container, he finally decided to give it a go.

And now his ride was coming. His skin began to sense the moment then he could just make his ride out through the droplets. This particular evening would be spent cruising town in Sara. Sara the socialiser, he thought, but promised himself to not repeat.

Sara pulled millimetrically into the blue parking bay.

"Is Hiroshi ready for some fun"? was called out in an unexpectedly loud female voice.

Sara had already detected him, at 92% probability, but was designed to double-check identities through conversation and the occasional well-disguised interrogation.

"Sara is here for Hiroshi. Please come to the corner where it is safe to pick you up", it said while changing skin colour from azure to amber, rainbowing its impatience at the delay.

Once through the portal-like door and absorbed inside the cabin two forms, one at right and one at left, proved to be people. People which appeared to Hiroshi to be roughly his age.

"Hello", said one.

"Cheers", the other.

"Pleased to meet you" was Hiroshi's by now well-practised English reply.

The people forms proved cordial and personable and the evening seemed off to a good enough start. I've never been on a social shuttle before, he thought, but so far so good. Sara estimated that the social interactions were already reaching an intensity of 52% which was not bad for so few minutes of concourse.

Social shuttles had become something of a fad of late. Lots of Londoners were riding them. Over the years it had become harder and harder to meet people and it was simply no longer possible to seek friendship from the shadowy images which passed for work colleagues. Or to place one's destiny in the hands of chance encounters. Social shuttles were now "the thing".

But they were also prime real-estate for those with opinion to voice. The influencers and the anti-influencers were battling for every metre of trench in terrains ranging from the environment to the local football team. The latest trends of "socials", "mixers", "speed dating" and other such encounters were great but could also be risky. There was nothing worse than being ambushed by a sales rep or a political activist when one was trying to have a fun night out.

But this social shuttle and these people forms seemed promising enough. The shuttle was azure in colour, pleasantly female in voice and seemed no stranger to a good joke (lots of shuttles were terrible at jokes unfortunately). With grippy-looking tyres, nice seats, mini-tables and accommodating interior décor. A pleasant and upscale shuttle, Hiroshi thought, as he turned his attention to analysing his fellow socialisers.

"Hi, I'm Alessandro", said one.

And "I'm Liz".

Alessandro was tall and thin and had an air of programmer to him. Indeed, Sara estimated his profession as being in IT at greater than 87% probability. His eyes moved about to focus on many points of interest but rarely met those of his fellow passengers. And his fingers seemed continuously engaged in some extremely urgent coding, clicking away against surfaces which had no intention of budging. Liz was instead shorter and stockier and harder to typecast. Maybe a local. Maybe not. Maybe someone who earned her keep doing something creative. Eyes like heat-seeking missiles which could not be shaken off. Sara estimated graphic artist at 42% probability.

It was still just after 18.00 thus the early birds were the first of the seven people which Sara would likely pick up from various locations around the city in the coming hours. The science says that socialising is best in seven. Thus Sara picks them up. Even if a few extra loops might be needed south of the river or an occasional romp over rutted Soho alleyways.

As they warmed to their circumstances Sara too began to enter into the thick of things, deploying social support capabilities. Questions about music selection were asked in 20% of Sara's interactions and the reactions to the probes of music taste were measured. Puzzles, games and trivia were rolled out on internal screens in 15% of the interactions as Sara tested the passengers in search of the optimal "fun point" on the standard chart. Observing Hiroshi's restraint, the shuttle prodded him to talk about tried-and-true topics such as his hometown and his reason for coming to London. Sara's "experience index" was soon hovering around 7.8, a pretty good result with first-timers. Everyone knew that more than 9.2 was nearly impossible for any shuttle to reach, even with the most extroverted passengers.

Sara-like social shuttles had already been in circulation for a few years, moving people around the city in the manner of a tour. At first, the shuttles were rather simple affairs, with a few items of pre-programmed tourist information and a few ice-breaking games to get things rolling. But, of course, things got more complicated. They always get more complicated. Sara was in fact one of the newest in the fleet and a rather sophisticated one at that. Sara could learn people's preferences, squirreling the information away. And could archetype a person's behaviour, estimating what might be said or requested next. After months of service Sara had by now, like a master blacksmith, fully learned the trade. Sensors had noted nooks and crannies all through the fabric of the city which were guaranteed to stimulate conversation, and the on-board memory was now chock full of traffic and weather conditions which fed into the pre-programmed socialising systems. Sara knew how to get people talking. And more.

By the time Sara reached the favourite haunt of Battersea Bridge the passengers were jovial and in each other's confidence. There was now talk of the fall from grace of the local sporting hero and about the wild weather of recent months. And there were the usual debates over English food and the friendliness, or lack of, which characterised the locals. Barriers had tumbled and walls had been holed. The décor, scenery and social support systems had worked their magic. Sara estimated a probability of greater than 90% of a successful evening and a 66% probability that one or more friendships would ensue from the evening's activities. At less than half way through the evening Sara's social sauce had proven a revelation.

But, then, a politics warning began blinking across several of Sara's internal systems. At first slowly and erratically, then with increasing frequency and intensity. A human being would have thought that it was essential to keep the passengers from getting on to politics or the evening would be shot. And in its own Westworld way, Sara must have been thinking something similar. Sara worked hard, repeatedly, to nudge the passengers off subjects which correlated with political discourse according to latest corpora. And for a time the shuttle succeeded at keeping the political content at less than 4% and the politics warning blinking only intermittently.

Then, when skirting Green Park and slipping past the Japanese embassy, Alessandro asked Hiroshi what he thought about the Asian League decision which had filled the media with such foul mouthed streams of consciousness in recent weeks. Hiroshi paused, for much longer than he would have wished. What do I say? And Sara's electrons sloshed to and fro to think, what do I do?

Big Easy Funerals

In most places it was proving rather difficult to use friendly neighbour-hood robots as funeral hearses. Heck, it was difficult enough to get them to do pickups and offer small talk well enough to avoid pissing people off. Yes, robots were now nearly everywhere, running around as taxis, delivery vehicles, road repairers, rubbish collectors and myriad other jobs which people seemed more than happy to hand over to them. But funeral hearses? There was hardly a single one in the country. Never mind in a venerable old town of time-honoured traditions like New Orleans.

And the funerals in New Orleans were different. Everyone knew that. Even the old-timers who sat half their life away glued to a chair grasping a data feed would know that in their bones. Yes, there were some sad affairs involving short hops to the crematorium or one of the few remaining ground plots, and yes there was the odd ritual involving elaborate religious customs and depressing eulogies. Many funerals, however, involved New Orleans jazz.

They might start out sombre, with slow and deliberate walking, as if the universe itself had come to an end and everyone was heading for eternal storage. But at the appointed time they would burst into music and begin shuffling, as if purifying the air and sweeping out the evil spirits. Dance would often break out and onlookers would join in, even if the deceased was their local taxman or neighbour from hell. A show if ever there was one. Sending the dearly departed off to their creator with a bang was the New Orleans way.

But a friendly neighbourhood robot as a funeral hearse? And, yet, there they were. Big Easy Funerals on Washington Avenue, Gert Town, now had their first friendly neighbourhood hearse. And today was the dry run. No grieving mourners, only a few staff, just to check that that it did not run over anyone or drive off into the river.

Daryl had insisted on the purchase. As a leading member of the local community and proud provider of funerary services he had felt it his civic duty to do so. From his casket warehouse to his family shuttles he had introduced the Big Easy to the art of funerary automation. Plunging into the brave new world. Telling everyone, but above all himself, that "It's the future".

The ancient and time-honoured explanation had worked for a while. But now the emotions were rubbing and sliding and poking from within. Breath shortened, lips curled and body parts moved about with their own agendas as he struggled to keep a confident face. God the cost of the thing. Weeks of training. A maintenance schedule which would put NASA to shame. And if it dropped someone's grandmother?

And many of the staff of Big Easy Funerals were worried. Would robot face manage to keep to the plan which was explained to it? Would it realise if it was misunderstanding a street name? Would it engage the right text sequences in response to questions from mourners? Would the robot adjust its speed, lighting and sounds to match the circumstances? To the crew, it just seemed like too many things could wrong.

"We're not transporting crabs or catfish with this thing", someone complained from behind.

"I bet it starts blasting out jazz during the priest's hail Marys", murmured someone else.

Though a somewhat superstitious gesture considering the line of work, it seemed that everyone had crossed their fingers.

The empty casket was loaded, doors were shut manually, and off the group went down the street. Slowly at first, then picking up speed a little. Initially the robo-hearse matched the humans nearly step-by-step, maintaining speed and achieving sobriety of its lighting and moving bits. A little misunderstanding occurred when they halted to let a human catch up and join the group, but, otherwise, all ok. Daryl, at the back, felt his chest becoming just a little less tight.

As planned they made their left turn into Lopez Street. Again, robo-hearse kept formation, kept mood and kept quiet. And as they approached the point of the procession where the music should start, robo-hearse opened up at just the right moment, emanating sound in all directions, shaking walls and moving hearts. Harmonies and melodies flowed from the robot, which to all the world seemed to have come from a complete orchestra of humans, brass section and all. The decibels discharged down the street and into the alleyways.

Then, however, something quite unexpected happened. Perhaps mistaking its own acoustic outburst for a police siren or ambulance, robo-hearse veered right and drove up the sidewalk, ordering, "Out of the way please this is an emergency". Fingers instantaneously uncrossed, chests tightened and a dozen balloon-like eyes chased down the street after the robot.

Robo-hearse eventually came to a dead stop in front of a fried chicken outlet. And as customers gathered round it began demanding they show respect for the dead man's family. As the minutes passed the stares only increased robo-hearses' determination. "Please be quiet", "Please do not disturb the procession", "Please do not force me to contact the authorities" and on went the avalanche which was growing in both volume and menace. A tired-eyed, grey-haired grandmother dropped her food and began repeating the signs of the cross with wide energetic gestures. Several people panicked as robo-hearse demanded they approach the casket.

The initial shock gradually slid down into the valley of disappointment for Daryl and crew. They had, they thought, followed the manufacturer's instructions to the letter. They had, they thought, explained all the movements, chosen all the verbal exchanges and set the preferred moods at least a couple of times, maybe more. All the items of the pre-funeral checklist seemed in order. All the key scripted scenes had been practised beforehand. Robo-hearse's body and mind, so to speak, should have been ready for the big performance.

"Maybe there's a reason robo-taxis are everywhere but you can't find a robo-hearse if your life depends on it", said one of the crew, sarcastically.

"Ready for New York but not for New Orleans", responded Daryl more determined than resigned.

Admittedly, not all was lost. It did many things right. Maybe there was just a problem with the sound sensing or with the self-identification. Maybe, just maybe, with more work, robo-hearse could be made to cruise the streets with confidence.

References

Abe, R. 2019, Introducing autonomous buses and taxis: quantifying the potential benefits in Japanese transportation systems, *Transportation Research Part A: Policy and Practice*, Vol. 126, pp. 94–113.

Bucchiarone, A., Battisti, S., Marconi, A., Maldacea, R. and Ponce, D.C. 2020, Autonomous Shuttle-As-A-Service (ASaaS): challenges, opportunities, and social implications, *IEEE Transactions On Intelligent Transportation Systems*, Vol. 22, No. 6, pp. 3790–3799.

Diels, C. and Bos, J.E. 2016, Self-driving carsickness, *Applied Ergonomics*, Vol. 53, pp. 374–382.

Hitti, N. 2019, Toyota redesigns its E-Palette vehicle for Tokyo 2020 Olympic athletes, *Dezeen Magazine*, https://www.dezeen.com/2019/10/14/toyota-e-palette-tokyo-2020-olympics.

Iskander, J., Attia, M., Saleh, K., Nahavandi, D., Abobakr, A., Mohamed, S., Asadi, H., Khosravi, A., Lim, C.P. and Hossny, M. 2019, From car sickness to autonomous car sickness: a review, *Transportation Research Part F: traffic psychology and behaviour*, Vol. 62, pp. 716–726.

Kim, S., Chang, J.J.E., Park, H.H., Song, S.U., Cha, C.B., Kim, J.W. and Kang, N. 2020, Autonomous taxi service design and user experience, *International Journal of Human–Computer Interaction*, Vol. 36, No. 5, pp. 429–448.

Kipp, M., Bubb, I., Schwiebacher, J., Schockenhoff, F., Koenig, A. and Bengler, K. 2020, Requirements for an autonomous taxi and a resulting interior concept, in International Conference on Human–Computer Interaction, Copenhagen, Denmark, 19–24 July, Springer, pp. 374–381.

Lim, Y., Giacomin, J. and Nickpour, F. 2021, What is psychosocially inclusive design? A definition with constructs, *The Design Journal*, Vol. 24, No. 1, pp. 5–28.

Lutin, J.M. 2018, Not if, but when: autonomous driving and the future of transit, *Journal of Public Transportation*, Vol. 21, No. 1, pp. 92–103.

Marchese, K. 2018, Renault's EZ-Ultimo car concept is a self-driving luxury lounge, *Designboom Magazine*, https://www.designboom.com/technology/renault-ez-ultimo-self-driving-concept-10-03-2018.

McArdle, M. 2022, The weirdest thing about being in a self-driving car, *The Washington Post*, https://www.washingtonpost.com/opinions/2022/10/31/waymo-self-driving-car-ride.

Nanchen, B., Ramseyer, R., Grèzes, S., Wyer, M., Gervaix, A., Juon, D. and Fragnière, E. 2022, Perceptions of people with special needs regarding autonomous vehicles and implication on the design of mobility as a service to foster social inclusion, *Frontiers in Human Dynamics*, Vol. 3, Article 751258, pp. 1–12.

Ortwein, O. 2015, Bookmobiles in America: an illustrated history, CreateSpace Independent Publishing Platform.

Riggs, W. and Pande, A. 2021, Gaps and opportunities in accessibility policy for autonomous vehicles, Project 2106 Final Report, Mineta Transportation Institute, San José State University, San José, California, USA.

Standage, T. 2018, What's it like to ride in a self-driving car?, Medium, *The Economist*, https://medium.economist.com/whats-it-like-to-ride-in-a-self-driving-car-24015c69cc48.

Turner, T. 2022, A hotel room on wheels, YD Yanko Design, https://www.yankodesign.com/2018/10/17/a-hotel-room-on-wheels.

Wikipedia Contributors 2023a, Bookmobile, Wikipedia, The Free Encyclopedia, https://en.wikipedia.org/wiki/Bookmobile.

Wikipedia Contributors 2023b, Robotaxi, Wikipedia, The Free Encyclopedia, https://en.wikipedia.org/w/index.php?title=Robotaxi&oldid=1128921610.

Yousfi, E. and Métayer, N. 2022, Improving mobility for people living with a disability: automated vehicles' opportunities and future challenges, SSRN, https://ssrn.com/abstract=4250791.

Chapter 7

Providing Workplaces

Background

Transporting people safely is the current focus of most autonomous road vehicle research. The achievement of safe, reliable and cost-effective transport is a major technical challenge characterised by far more complexity than imagined by the 20th-century pioneers of the sector. Interpreting surroundings, tracking moving objects, navigating a continuously changing cityscape, and interacting with pedestrians and other road users in legally and socially acceptable ways are non-trivial tasks.

But the successful achievement of safe and cost-effective driving is already being assumed to some extent (see, for example, Herrmann et al. 2018). The technical hurdles are expected to be overcome and autonomous road vehicles which can drive safely in rain or snow, communicate efficiently with people, and provide a range of capabilities and affordances are expected to arrive.

Thoughts are thus already turning to the services which the friendly neighbourhood robots might provide. Once safe and reliable autonomous road vehicle platforms are achieved, the creativity and curiosity inevitably turns to the question of what else they might do for people beyond transporting them. With automation advancing in capability and spreading in application at a rapid rate, there is a natural tendency to seek new ways of applying the new forms of automation to time intensive or cost intensive tasks. Particularly those tasks which humans find tedious, difficult, uncomfortable or unpleasant.

Already today there are a few autonomous road vehicle concepts which are more about applying service automation than about safe and reliable mobility. Several proposals and a few early experiments have investigated workplace support, healthcare or leisure activities. In these cases the focus has been more the service activity than the getting from point A to point B. Such proposals are more about what an autonomous road vehicle can do for people than how it drives or where it goes.

DOI: 10.4324/9781032724232-7

One line of reasoning involves what can be called "autonomous offices" or "robot workplaces". The premise is to make better use of travel to and from meetings by providing a supportive work-orientated environment. Or, alternatively, to provide a temporary office which can be called upon as demand dictates, avoiding the need for a bricks-and-mortar office. A friendly neighbourhood robot of this type can be considered a technological response to the spending of more time working outside of the traditional office. Such proposals attempt to close some of the gaps in workplace support which can occur with current taxi or mass transit services. They provide enhanced travel time efficiencies and opportunities for forming temporary facilities for site-specific projects such as construction sites.

There is of course an air of "incremental" to such offerings. Autonomous offices are a further evolutionary step from the best-equipped shuttle buses and train services of today. Many business-friendly transport services already equip their human-driven vehicles with electric power sockets, internet connections, computers and other items of equipment which are typical of office environments. Exploiting the packaging freedoms and cost savings of autonomous road vehicles to achieve a space filled with a large number and variety of communications links, information technologies and specialist scientific tools seems an obvious opportunity.

One example of the approach is found in the research study of Li et al. (2022) who investigated the use of virtual reality systems for performing office work within autonomous road vehicles. Cabin space and road vibrations permitting, the supporting equipment is integrated into the fabric of the autonomous road vehicle to provide a relatively self-contained office environment. Adding data communications links, information technologies or specialist scientific tools as they become available seems a likely destiny of the friendly neighbourhood robots.

And several research studies have pointed to the benefits of such approaches. A typical example being that of Stevens et al. (2019) which reported that passengers preferred dedicating their travel time to office work. Reading, writing, messaging, preparing presentations, sorting documents and talking to other employees were all activities which were specifically mentioned. And the Stevens et al. (2019) study was one of a number of investigations which also suggested the benefit of two different vehicle architectures for the future autonomous offices: special purpose vehicles for specific tasks and generalist robots which the researchers referred to as mini-houses.

Typical of the many autonomous office proposals is the "Work On Wheels" concept of the design firm IDEO (Bryant 2014). The "WOW" consists of interlocking cube-like offices on wheels which can be used on the road or used in parked formations to form temporary facilities. Besides seating and information technology support, the WOW concept also foresees smart desks, video projectors and other advanced forms of networking and co-working which would allow the single pod to act as the hub of larger meetings or of complete projects. And beyond supporting office activities, the IDEO designers emphasise that the concept also provides access to daylight, natural ventilation, views and green spaces.

Such friendly neighbourhood robots are likely to prove beneficial in areas characterised by high rental costs in buildings, where the office space is prohibitively expensive when not in continuous use. A rentable mobile alternative may prove attractive in such cases. And while working from home has proven popular in recent years, there is probably still business demand for mobile office facilities.

Another line of reasoning which crops up regularly in the specialist press is what can be called "autonomous workshops" or "micro factories on wheels". Less immediate, and more long term, the concept nevertheless appears a natural progression from current developments in factory automation.

Several experiments in factory automation have attempted to achieve autonomous robots or autonomous workstations which move from place to place as needed within industrial settings. For example, Ford (Hanaphy 2022) has tested an autonomous 3D printer which liaises with the other factory robots and moves autonomously to any needed location to manufacture any needed part. Such advances can be considered a step forward in machine automation along the path of achieving human-like flexibility and human-like decision-making.

In light of such developments there has been much speculation in recent years about the benefits of micro factories (see, for example, Industrial Equipment Blog 2020 and Sharp 2022). Micro factories have been claimed to be small, highly automated facilities which provide the flexibility to manufacture different artefacts at different locations in real time as needs evolve. Micro factories are claimed to be a natural response to the imperative of localising and personalising the offerings of large organisations, and have been claimed to be an essential support for future innovation.

Combining the concept of autonomous road vehicle with that of micro factory, and projecting forward to a time when the automation will be substantially more powerful than today, one arrives at the concept of "autonomous workshop" or "micro factory on wheels". Such proposals envision an autonomous road vehicle which is equipped with highly automated small-scale manufacturing equipment such as computers, paper printers, 3D printers and small injection moulding machines.

The reasons for developing a friendly neighbourhood robot of this type have been argued to include the ability to manufacture prototypes at remote sites or while on the way to meetings, the flexibility to boost local production when market demand shifts unexpectedly and the economic convenience of a rentable resource for entrepreneurs and small businesses.

There is more of an air of "disruptive" to such offerings because it is difficult to identify many current human-driven road vehicles which fulfil a similar role. The metaphor of "autonomous workshop" or "micro factory on wheels" would thus seem to be relatively new and untested, thus in need of much development work and much usage experience in the field. Nevertheless, once safe and trustworthy autonomous driving is reached, "autonomous workshops" or "micro factories on wheels" seem reasonably likely developments.

Such friendly neighbourhood robots are likely to prove particularly beneficial to innovators in creative disciplines characterised by either a rapidly changing technical context or low production volumes. Such a vehicle might also prove a helpful support towards co-design, co-creation and crowdsourcing since in such cases there may be no guarantee that the needed prototyping facilities are available to the contributors before-hand. As such, friendly neighbourhood robots of this type may have a role to play in the further democratising of innovation.

Autonomous offices, robot workplaces, autonomous workshops and micro factories on wheels are all forms of friendly neighbourhood robot which seem somewhat likely to appear on roads in the coming years. Such developments appear to be likely consequences of the current trends towards increased cost of brick-and-mortar facilities, increased need for flexible working practices and the greater uncertainties in the international markets.

What follows on the remaining pages of this chapter are a set of four science fiction prototypes, short immersive vignettes which explore the nature and implications of the future friendly neighbourhood robots. Each involves physical, psychological and social interactions with the future friendly neighbourhood robot and several allude to anticipated ethical conundrums.

As discussed earlier in this book, the science fiction prototypes are linguistically based and avoid an excessive focus on the robot itself, so as to leave ample room for imagination and speculation. It is the author's hope that they can provide a focal point for discussion of autonomous road vehicle design.

Dave's Diary

February 6th.

Today was a rotten day. As rotten as it gets. And only now in late evening am I managing to control the jumpys and sit long enough to write. Already last year I knew that there would be days like this, and now I've had one.

I knew that I was not going to like being in the field. As soon as I signed the contract the stomach burning and churning started and the feelings of nausea emerged from somewhere. My head started splitting every time I visualised one of the things. As I wrote last year, the lousy feelings lasted for days and drove home the fact that I was not going to like being in the field. But I needed the money.

I was happy as a programmer. And good at it. They needed me to keep that meandering mess that they call the routing code under control. Everyone knew it. Martin and his band of self-professed geniuses never stopped reminding me that I was the best programmer they had. So why did they think that I would be good in the field? Platooning lorries is not the same as ordering algorithms around. I told them so.

Sitting in the tiny cabin at front for hours and hours is not so bad. Sometimes it's quiet enough to catch up with my programming or kill some time entangling. And, yes, with eight or ten lorries in platoon the dealing with the occasional roadside repair, emergency or traffic jam needs someone organised and someone who can follow instructions. That's true. But I told them that someone on the spectrum needed more structure. I told them so. More than once.

And today I had my first tyre blowout.

The rotten rutted road blew the front right tyre to pieces with a loud bang.

Pavement littered with bits of carcass for a hundred metres or more.

Ten lorries stopped along the shoulder, losing time, waiting for the repair.

Waiting for me.

A mess.

Why do days like this lead to headaches which split the head in two? Why not three or four or five? Why always two?

And of course I applied the training and followed the procedure and broke the convoy, sending most of the vehicles off to destination individually. Thank goodness that they are able to drive themselves independently in emergencies.

But there was so much to sort out. And a lorry is not a computer. It's a filthy, messy, confused thing that has more in common with steam engines than with entanglers. A relic of a bygone age, a necessary evil.

And practically none of it has a hierarchy or an ordering or anything rational. Programming is pure, it has rules, and it either works or it doesn't. But lorries require twists and kicks and chemicals and all manner of dirty tricks to get them to work. God how I hate the things.

Was it really that hard to buy a more rational model than the one we use?

Was it really that expensive?

Why not use one that has labels, buttons and levers in clear view and always in the same places?

And why don't they equip the damn things with automatic tyre changers?

Why does the damn thing sort some matters out on its own but instead drop others in my lap?

I spoke with headquarters and followed their instructions to the letter, as ordered. Go to this, turn that, lift up that other, push this … They sent me to one side of the thing, then sent me back, then sent me to the one side again. Who designed the thing? And who trained the idiots at headquarters?

The roadside pushing and shoving and talking and negotiating ran right over me. Stomping. As if the cattle in the field just next to the road had run over my head rather than chew and smile. After a while I was stuttering again, shaking and having trouble stopping myself from smashing something. A proper case of the jumpys.

Tomorrow I'm going to give some thought to what happened today. I'm going to give some thought to why I was jumpy all afternoon, and try to work my way through a few issues in the usual way. Tomorrow I'm going to try to get to the bottom of why the blown tyre was such a rotten experience. Tomorrow I'm going to figure out some ways of keeping the jumpy under control. And I'm going to take the ideas to Martin and his geniuses.

And of course I'm going to tell Martin's pack that they should replace the chaotic smelly relics with something tidier. Something more organised. I know that there a few lorries with roomier cabins and with their bits and pieces laid out where you expect them to be, and labelled. A lorry is not a computer, I know that, but at least give us something that doesn't strain the brain. Give me something simple and precise to work with. Something which takes care of itself for the most part without needing me. Something that makes obvious exactly what it needs from me, when it does end up needing me. Give me something tidy and organised and I will sort whatever needs sorting.

I need things to be things, rules to be rules and jobs to be jobs.

That's all I want.

And if they can't provide me with something sensible and logical to work with, then screw them.

The money isn't worth it.

Plumber Jim

The colleague had just pulled up in front and he was eager to conclude the business, if possible. He had decided that he would trade his vehicle in for a larger one given the growth in his client base, and had started making plans to advertise the machine. But when his long-time colleague appeared interested he grudgingly agreed to offer it to him first. Seemed the right thing to do.

"Welcome, Pete, good to see you again" was the opening shot over the fence.

"Thanks for finding the time to show me the machine" was the return volley from the sidewalk.

"I've been needing one for some time and maybe today is the day".

"Shall we have a look"?

Jim, a plumber from an early age, had by now lost count of how many sinks and showers and taps and traps he had installed. He was good with people and better with the tools of the trade. He was respected among the members of the piping public and his opinions mattered. His vehicle was almost certainly a good one.

A gravel driveway passed underfoot, a side of the house came and went, and the two eventually appeared in front of a small garage at back. A large blue vehicle with diverse doors and hatches sat inside. A large blue vehicle which had been Jim's companion of many years. A chauffeur, piper-fitter and jack of all trades.

"As you can see, Pete, the machine might be a few years old now but it is still nearly as it came from the factory. Nothing but a couple of scratches on the left hand side, and nothing not working like it should. A wash and a couple of touch-ups and nobody will know that it is not straight from the factory.

"And you can see that this one is a full installation model. It has all the space needed for the usual stuff. It has its own pipe cutter, bender, solder machine and welder. And at the back the fold-out has a woodworking bench, saw and the other tools we use with wood and plastic. Lots of room and lots of kit. You don't have to rely on anyone when you have one of these.

"And of course the connectivity suite and pickup feeds inside provide everything you need to access the municipal sewer plans, track down component drawings, order parts and do other money-making things while on the way. Heck, if it weren't for your wife and kids you could just sleep in the thing and live your life out of it without the need for a bricks and mortar".

"What about when the usual jerks who try to nick the kit, Jim, or try to nick the machine? What does it have to defend itself"?

"The usual. The alarms, the automatic owner contacting and the road certification to drive itself to safety if needed. They all have those things since the authorities passed that robot rights amendment and made messing with these fellows a criminal offence".

"Ohhh I see. I didn't know that. I thought you had to pay extra. Shows you how much I know"…

"What about heavy-duty water pumps or high water suspension lifts for dealing with flooded homes or burst mains? Does it have any of that stuff"?

"Sorry, Pete, can't help you there. I've never actually needed them. But, I think I remember that the C panel came from the factory with the power and control sockets for fitting the pump. And I'm pretty sure that the suspension lifts can be installed. There is an indicator for the suspension height which pops up when you go to manual. I don't think it would be there if it was something which the machine couldn't do. I'm pretty sure that can you drop in the lifters the next time you get it serviced.

"What do you think, Pete? Can you use the machine or is it not what you expected"?

"No, seems about right. Just a big lump of money, you know, not something you buy every day. And with my boy heading off to uni and bleeding my bank account dryer than a vampire's victim, it's just not an easy thing to do".

"Fair point, but I do need to move the thing and get on with tracking down a bigger one before the holiday. And of course I consider you a friend, Pete, and don't want to sell it to someone else if the thing is right for you. You know that you can trust me. My parents were bent on raising me straight.

"I tell you what, Pete, I've been using one of those avatar projectors for some years now to deal with the punters while I get on with fixing their pipework. It's great for dealing with people and their pets. God knows the amount of time that some people can lose you. It's been a godsend, keeping them busy while I get on with matters. I'm pretty sure that the system connectors on the new vehicles have changed and are different from the projector I've got, thus it's likely that the projector won't work with anything which I might scrounge up. Is the projector of any use to you? I could throw it in with the vehicle to swing you and close the deal.

"What do you say, Pete?

"Have we got a deal? Or are you going to leave here like a dope and rush over to one of those equipment sharks who will rob you blind, selling you a machine which can't even drive itself without running over lawn gnomes and driving into swimming pools? What do you say, have we got a deal"?

"Ok, Jim, you've got me".

"I need one, and I might as well pay for your holiday rather than send some greedy bastard to the Bahamas".

"Ok, then, that's that".

"You won't regret it".

"And if anything does go pear shaped, just call me, and I will get my finger out to help you sort it".

"Now let's get down to the Rose and Crown – first round's on you, Pete".

HFT1

Financial markets had been turbulent lately with bulls following bears and bears following bulls, a zoo that needed navigating without a map of the enclosures. A mess even by the standards of mess. And nerve-wracking even for the people who had made their first million before they had ridden their first bicycle.

Sir John, as he was known to his peers, was now rushing to Biggin Hill for his 20.25 flight to Brussels. One of a flurry of such trips precipitated by the market turbulence of the last week. During quieter periods the need for such travel would have been once a week at most, but at the moment the conveyor belt was stuck at the "on" position.

On this occasion High Frequency Trader 1 was Sir John's ride to the airport. A top of the line trader shuttle. Hired due to the need for a rescue craft to ride out the market storm. Deploying HFT1 did not guarantee success, but it helped. The right tool for the job. A safe port in the storm.

HST1 was certainly well equipped. Enough data screens streaming enough live data to make a traffic controller jump for joy. And enough computer power to permit high-speed trading from a moving platform, a feat of engineering which up to a few years prior would have been considered science fiction. And to cap things off, a three-dimensional avatar controller to permit up to six different Sir Johns to attend six different meetings anywhere in the world, all under Sir John's real-time supervision. Sir John could buy up half the planet and sell it back again from the cocooning comfort of a soft sofa while racing to Biggin hill at 60 miles per hour. Definitely a great ride if you can afford it.

And today HFT1 was earning its none-too-economic rental cost. Sir John was in three different meetings before even reaching Blackheath. And had a couple of algorithms attempting to achieve early morning latency arbitrage on Asian equities posting out of Shanghai. Navigating the road was none-too-bothersome for HFT1 but handling all the telecommunications traffic was keeping it very busy indeed. All the hot silicon was coaxing the cabin temperature from comfortable to steam sauna.

By 19.30 the first arbitrage was accumulating. Not enough to plug the outflow holes, but good nonetheless. Enough to bring a quarter smile to Sir John's lips despite his professed dislike of any form of manifestation of any form of pleasure. And by 19.45 the first meeting had concluded with an agreement between all the partners. Another satisfactory result and another good omen.

At 19.55, however, a snag occurred. A gap in the telecommunications coverage near Bromley led to a suspension of the high-frequency trading and to glitches in the parallel virtual Sir Johns. One of the holograms, in particular, had agreed a price point which was well outside its settings and had even moved to sign the contract despite the limit set for its avatar engine. Much money was about to be lost on account of a lethargic link. Hologram Sir John was behaving like someone who had enjoyed one too many at the skylight lounge.

As the glitches became ever more apparent and the behavioural errors were tracked back to their source, the biological Sir John was informed of the badly behaving second virtual Sir John. HFT1's control circuits bristled with activity and from somewhere deep in the depths of its silicon soul it politely proposed a recommended remedial action to bring biological Sir John and virtual Sir John back into harmonious accord.

Three, two, one, engage new instructions. Upon biological Sir John's command, second virtual Sir John excused itself from its meeting to liaise with base, and then re-entered as ordered and brought the discussion back to the problematic price point which had been proving the bone of contention. Reason and realism had been restored. Within minutes, second virtual was back on track and behaving like a sober Sir John again.

By 20.02 the third virtual Sir John had concluded its meeting successfully, and the second virtual Sir John had restored the situation and achieved a compromise for the disputed price point. Topping off the run of successes was the welcome sight of a highlands-like roadway, bereft of traffic. Biological Sir John was now nearly in front of the airport gate and guaranteed to catch the flight on time. The afternoon had seen success wash up everywhere despite the morning having promised only a skeleton coast.

And at this very moment, 20.02 precisely, unequivocally and verifiably, a three-quarter smile emerged on Sir John's lips despite his professed dislike of any form of manifestation of any form of pleasure.

Marrakesh Making

For someone who feared flying as much as he did this was about as good as it gets. Silky air, clear sky and stunning desert views stretching to the horizon. No terrors, no nails sunk into armrests and no praying to gods he did not believe existed. Two turbulence-free hours and the fuzzy contours of Marrakesh Menara Airport now materialising from behind the desert haze.

Breathing slowed, chest expanded and blood pressure normalised. The top button of his shirt was now once again in its place and his seat was reclining downwards at a leisurely pace. Thoughts could now replace fears and normal life could resume. And soon enough Erwin was launching into his absolute favourite pastime, planning. Proper planning. So much to do, so many decisions to take, so much to schedule, best to get at it right away. While still hundreds of metres in the air his thoughts were already now firmly on the ground.

Erwin was the founder and owner of a small electronic components company in his hometown of Dresden in Germany. At forty-two years of age he was still relatively young, at least by modern standards, and had been born sufficiently late to avoid both the horrors of war and most of the restrictions of life under the communists. Fascinated by electronics and computers from an early age, he was now expanding his business abroad via local design and manufacture collaborations. Fear of flying somewhat soured his trips, but what had to be done had to be done.

Fumbling to hold on to the coagulated clump of papers and permits, Erwin noticed the time. Forty minutes since exiting the gates and his taxi-office had not yet arrived. Forty minutes of entering and exiting the terminal, alternating between cryogenic freeze and furnace-like heat. Multiple calls to Germany and an avalanche of expletives later the crisis was still ongoing. Despite strenuous efforts his company simply could not get a reply from the taxi-office service. "Unavailable" was the stable state and a mild migraine was the unavoidable upshot.

Erwin's synapses were grumbling thoughts of hourly staff cost multiplied by lost time. Time is money, money is time, and Erwin was losing plenty of both. Need a taxi-office was the irritating left-brain thought which repeated with clock-like continuity. While desperate times demand desperate measures was the even more irritating right-brain response which chimed into consciousness at regular intervals.

Soon a heated exchange was under way with the young man at the tourism kiosk, one which increased in tempo, pitch and intensity as the minutes passed, colouring cheeks and darkening brows. "Please can you book a taxi-office for me"? was clearly discernible in the manner of a favourite harmony. And "needs to be well equipped" and "cost is not a problem" seemed to be providing background percussion. Finally, red turned to pink turned to tan, and success. A taxi-office was on its way. Synapses were standing down, the thought clock was ticking no more and the blood returned to Erwin's cheeks.

The taxi-office soon arrived at a blue square in front of the Marrakesh Menara. The only sound it seemed to be making was that of the air being brushed aside as it glided in. Good thing I have my eyes wide open, thought Erwin, or I would have no idea that it is here. Erwin had been told to look out for a yellow robot, but as it approached it seemed to him to be more dust beige in colour. Perhaps the desert heat and sand had worn the robot down, or perhaps the robot was smart enough to somehow mirror its surroundings. Either way, there it was.

A large sign announcing the company's name sat squarely on top of the roof, maybe to avoid people trying to flag it down between riads thinking it was a shuttle. The big bright board reminded Erwin of the New York taxis of the 1940s and of how the old ways can sometimes be the best ways. Mostly in English, it spelled out "Marrakesh Making" in wide wiggly letters. Something was also written is small Arabic script below, but Erwin had about as much chance of reading it as he had of translating hieroglyphics. Possibly the name of the company, or the type of robot, or maybe something else.

The robot was big. Nothing like the petite propositions which hummed down alleyways back home. Very wide, maybe too wide, for a city of narrow tight riads like this one. "Let's hope we don't get stuck in a souk", said Erwin without directing his comment to anyone in particular. Maybe the roof line was set high to accommodate the communications and manufacturing systems without cramping the cabin. Clever.

Once inside he marvelled at how the taxi-office now seemed even bigger than it had from the outside. Optical illusion? Secret compartments? Flight fatigue? He couldn't be sure. A feed panel protruded downwards from the roof, its video cameras, audio sensors and entangler poking into the space. Lower down were paper printers, 3D printers, a 3D scanner, a hologram projector and enough other goodies to satisfy even his designery demeanour. And contouring the edges of the workmanlike interior were cushions and upholsteries in brownish greys, yellowish reds and a cactus colour of unknown name. A forest of forms and scents which cuddled the workspace and exorcised the residues of inks, papers and plastics.

Inside he also had his first encounter with the taxi-office's personal assistant mode as it hurled questions at him. "Do you have a destination where you wish me to take you Mr Erwin"? was the first volley, followed in rapid fire succession by "Would you like me to connect you to anyone Mr Erwin"? and other pertinent probings. And before he had even completed half of his replies the Marrakesh Making was at speed and stretching its communications bandwidth to its limits.

Personal assistant mode soon prodded Erwin about his maker needs and offered various items of equipment in support. Whispering the name and gesturing upwards with his forefinger got his company's bundle connector projected into the air. A nod of the head in the direction of the 3D printer and an upturned thumb then got the production under way. Within minutes a satiny prototype was sitting on the stand and ready for the local subsidiary. A job well done. A few seconds of eye gaze directed at the words "Financial Times" on the cabin wall then conjured up a smoky trail of news items and stock summaries which scrolled from his left and seemed to bend around the front of the taxi-office, disappearing somewhere into its bowels.

Despite having argued with a tourist officer and waited excruciatingly long for a ride, Erwin was now back on schedule and exactly where his meticulously planned day should have been in the first place. Efficiency had been achieved. Satisfaction now oozed and peace descended upon the land. A great taxi-office this one, he thought. I wonder why we don't have ones this good back home?

References

Bryant, R. 2014, IDEO envisions a future where self-driving workplaces commute to you, *Dezeen* magazine, https://www.dezeen.com/2014/11/26/ideo-self-driving-driverless-vehicles-car-21st-century-mules-cody-truck-wow-pod-work.

Hanaphy, P. 2022, Ford rolls out autonomous robot-operated 3D printers in vehicle production, 3D Printing Industry magazine, March, https://3dprintingindustry.com/news/ford-rolls-out-autonomous-robot-operated-3d-printers-in-vehicle-production-206239.

Herrmann, A., Brenner, W. and Stadler, R. 2018, *Autonomous Driving: how the driverless revolution will change the world*, Emerald Group Publishing, Bingley, UK.

Industrial Equipment Blog 2020, Microfactories – the next big thing in manufacturing, FutureBridge, October, https://www.futurebridge.com/blog/microfactories-the-next-big-thing-in-manufacturing.

Li, J., Woik, L. and Butz, A. 2022, Designing mobile MR workspaces: effects of reality degree and spatial configuration during passenger productivity in HMDs, *Proceedings of the ACM on Human–Computer Interaction*, Vol. 6, Issue MHCI, pp. 1–17.

Sharp, N. 2022, Think global, build local; the rise of the micro-factory, ESCATEC, April, https://www.escatec.com/blog/think-global-build-local-the-rise-of-the-micro-factory.

Stevens, G., Bossauer, P., Vonholdt, S. and Pakusch, C. 2019, Using time and space efficiently in driverless cars: findings of a co-design study, in Proceedings of the 2019 CHI Conference on Human Factors in Computing Systems, Glasgow, Scotland, UK, 4–9 May, ACM, New York, New York, USA, pp. 1–14.

Chapter 8

Providing Healthcare

Background

As noted earlier in this book the safe transporting of people is the focus of most current autonomous road vehicle research. Achieving friendly neighbourhood robots which interpret surroundings, track moving objects, navigate a continuously changing cityscape, and interact with pedestrians and other road users in legally and socially acceptable ways is the current goal.

However, once safe and reliable autonomous driving is achieved, the creativity and curiosity will naturally turn to other things which the friendly neighbourhood robots can do for people. With automation advancing at a rapid rate there will be a natural tendency to seek ways of applying the new forms of automation to other time- or cost-intensive tasks.

One line of reasoning involves what can be called "contagion control robots". For example, Khalid et al. (2021) have suggested that

> ... use of autonomous transport services amid pandemics such as Covid-19 can offer improved emergency and healthcare services by limiting exposure, improving response time, offering better management of patient pickup and dropoff, and facilitating more patients with the same number of ambulances. It also allows effective delivery of medicine to homes without exposing patients or drivers to Covid-19.

And Huang et al. (2021) have emphasised that patient transport by means of autonomous road vehicles can help to protect the key medical workers from contagion.

"Contagion control robots" are minimally modified machines which can move people or transfer medical samples to emergency centres. Or transfer masks, medicines, foods or other goods to the ill individuals at their home. A contagion control robot is an easily

DOI: 10.4324/9781032724232-8

cleaned and easily decontaminated shuttle which can be deployed in a flexible manner to perform several different tasks, limiting the direct human-to-human contact.

Early experiments with such machines occurred during the Covid-19 pandemic when several simple shuttles were used to transport people or materials. For example, the Mayo Clinic in Florida used four NAVYA shuttles to transfer medical supplies and Covid-19 test samples from a drive-through testing facility to a nearby laboratory (Blanco 2020). And in China Neolix shuttles were used to deliver medical supplies in the city of Wuhan as well as to disinfect streets (Linder 2020). The small-scale experiments have produced hundreds or perhaps thousands of hours of usage data which is helping to refine the design specifications of the vehicles.

Another line of reasoning involves what can be called "autonomous ambulances" or "emergency response robots". For example, Elayan et al. (2021) have noted that the vehicle-to-everything (V2X) connectivity of many autonomous road vehicles allows them to optimise their navigation of the local traffic conditions to such a degree as to reduce the time to first treatment by up to 75% with respect to traditional human-driven road vehicles. Potential driving efficiencies alone have been suggested to be sufficient to justify the use of autonomous road vehicles as ambulances, even in the absence of additional medical capabilities.

"Autonomous ambulances" are specialist friendly neighbourhood robots which can transfer people to emergency medical centres or to hospital accident and emergency reception. They can be equipped with a range of sensors, monitors and simple life supports such as oxygen supply. Easily cleaned and easily decontaminated, they move patients rapidly from home or from the scene of an accident to the point of emergency medical care. Real-time data links are often envisioned to provide monitoring and intervention capabilities from remote locations for the medical staff who support the patient or patients during the journey.

Typical of the current proposals is the Volkswagen ID.Buzz electric ambulance (Cenizo 2021). Based on an existing electric autonomous road vehicle platform, a variety of medically relevant devices and services are placed within the confines of the standard vehicle. In the manner of traditional human driven ambulances, the vehicle platform is built up to meet the requirements of the ambulance role via the addition of relevant items of emergency care equipment. By adopting a traditional approach to ambulance design the ID.Buzz ambulance represents an incremental step forwards towards of a more highly automated future.

Early surveys of public reactions to autonomous ambulances have suggested challenges to the building of trust. For example, Winter et al.

(2018) noted that females were "less willing to ride in the automated ambulance than males" and that there was a "significant interaction between the ambulance configurations and willingness to ride". However current experiments are beginning to identify possible trust-building remedies involving the provision of real-time medical informa-tion and real-time interactions with artificial or human healthcare workers. Thus the establishment of trust in such friendly neighbourhood robots may not prove to be an insurmountable obstacle in the long term.

And a further and more sophisticated line of reasoning involves what can be called "autonomous clinics". Typical of this area of specialist discourse are the claims of Liu et al. (2022) who suggested that "Autonomous Mobile Clinics (AMCs) have the potential to solve the healthcare access problem by bringing healthcare services to the patient by the order of the patient's fingertips". The core concept is for friendly neighbourhood robots to provide one component of an integrated healthcare system which combines electronic health records, medical equipment integration, autonomous driving and telemedicine.

"Autonomous clinics" are friendly neighbourhood robots which are maximally medically equipped to provide a variety of testing and diagnostic services. Such robots should travel to people's residences, offices or leisure locations to provide scans, tests and diagnostics within the confines of the vehicle itself. Advantages of such an approach include convenience in the case of regular health check-ups and rapidity of medical diagnostic response in the case of emergencies. Easily cleaned, easily decontaminated and capable of disposing of test samples, such machines are envisioned to also provide real-time data links to medical staff at remote locations for purposes of oversight and consultation.

A few human-driven road vehicles are already providing early tests of the concept. For example, Bannerman (2023) has reported the experi-ences of the Physician Response Unit (PRU) which began with a doctor in a car providing ground support to the helicopter crews of the London Air Ambulance. The PRU currently consists of three cars which attend emergencies across northeast London, bringing a doctor, emergency medications and a mobile IT system which can access the patient's electronic health records. The service is claimed to bring A&E to the patient, and has managed to treat outside hospital about 70% of the patients who would otherwise have required ambulance transfer to hospital.

And an interesting proposal for a fully autonomous clinic is that of the "Aim". The Aim autonomous clinic (artefactgroup 2023) specification is suggested to include controlled lighting to facilitate visual assessments, optical and thermal imaging technologies, scales to measure mass, BMI, balance and posture, breath analysers, a seat capable of measuring respiratory and cardiac rhythms and AI-driven real-time instructions to the patient. Artefactgroup also claims that "Aim dispenses the most frequently needed pharmaceuticals, like prescription analgesics, antibiotics, or contraceptives" and that "If the self-guided assessment indicates the patient needs to consult a specialist, the Aim platform connects the user to one of the on-call specialists from participating fleet partners".

And, finally, it should be noted that future friendly neighbourhood robots will also offer opportunities for psychosocial support (see Lim et al. 2021 for details). It does not seem unreasonable to suggest that some robots will support not just physical transport needs but also some of the psychological and social needs of the passengers. Perhaps even from within a healthcare or social care setting. Given the increasingly blurred boundaries between healthcare, social care, leisure and lifestyle, there may prove to be a variety of current metaphors which, when mixed in appropriate proportions, might meet the psychosocial needs of a given group or specific community.

"Social support robots", for example, will be specialist friendly neighbourhood robots which can interact linguistically, emotionally and socially with their passengers and provide a degree of support, companionship or possibly even psychiatric therapy. Such robots will be able to travel to people's residences, offices or leisure locations to provide comfortable protected spaces. And may use their mobility to offer specific drives or tours of therapeutic relevance. Proposed advantages of such friendly neighbourhood robots include that they may prove more quickly available than human friends or family members and will benefit from medical databases and optimised algorithms which may prove better able to provide specialist support.

Contagion control robots, autonomous ambulances, emergency response robots, autonomous clinics and social support robots are all forms of friendly neighbourhood robot which seem somewhat likely to appear on roads in the coming years. Such developments are likely consequences of the current trends of increased lifespan, increased quality of life expectations, and increased medical costs which place greater demands on healthcare and social care systems.

What follows on the remaining pages of this chapter are a set of four science fiction prototypes, short immersive vignettes which explore the nature and implications of the future friendly neighbourhood robots. Each involves physical, psychological and social interactions with the future friendly neighbourhood robot and several allude to anticipated ethical conundrums.

As discussed earlier in this book the science fiction prototypes are linguistically based and avoid an excessive focus on the robot itself, so as to leave ample room for imagination and speculation. It is the author's hope that they can provide a focal point for discussion of autonomous road vehicle design.

Tovar Tester

The clinics neighbouring the Plaza Bolivar and Universidad Central guaranteed that there was no such problem in the capital, but in the small towns it was a different story. In the towns the locals were lucky to find a doctor, any doctor, never mind somewhere to test blood or scan a kidney. In the towns the hunt for medical help usually cost an arm and a leg, sometimes quite literally. And reminding the health ministry of the need to make a good impression on the tourists had not swayed them in the slightest.

Colonia Tovar is less than 40 km from the heart of the capital but it might as well be on the other side of the continent. Twenty thousand potential patients with only a handful of doctors to care for them. And a couple of leaky clinics whittled down by tropical storms, now more barns than buildings, where water and aspirin were often the only cutting-edge technologies. Sometimes when conditions conspired and supplies shifted and people persevered, tests of blood or urine or faeces could be achieved.

The minimal medical service had long been a source of irritation for the locals and a bone of contention with government officials. Custom was for the townspeople to make the bumpy two-hour journey down to the capital if they needed a doctor for anything more than a cut or bruise or minor infection. Wooden seats, packed bodies, heat and a sidewinder road were the problematic particulars. And life-threatening accidents were far too frequent. But custom yielded when the tester finally arrived in Colonia Tovar.

The tester, which the locals quickly nicknamed the "Tovar Tester", was one of the friendly neighbourhood robots which provided level 7 medical support. They were white, with large red crosses on all sides, signalling lights and screens from end to end. They could automatically collect sweat, saliva, urine, faeces or blood for testing, providing results within minutes, and incinerating the sample immediately afterwards. All testers could perform most routine medical checks on the spot. The main exception being the recently introduced pan-RNA-screening methods.

Given the delicate nature of their activities the testers were inten-tionally simple, so as to cut down on confusion. They responded mainly through buttons and a tiny vocabulary of human words. Uttering "urine" and "infection" initiated a urological test while blabbing "blood" and "iron" brought haematology into play. The medical equivalent of a soda machine.

Two main types were used in the country. "Standard" was much like most friendly neighbourhood robots, four run-of-the-mill wheels and a large boxy cabin. "Remote location" was instead a more expensive proposition. An all-wheel drive device with more navigation algorithms, more weather prediction, a road-surface tester and other safeguards to ensure that it did indeed reach its remote location. Eight large grippy wheels sat below a tall and somewhat sloped cabin. Like the "standard" it had all the crosses and lights and other gadgets, but the rumbling powertrain and crunching tyres suggested a different DNA. A godsend for those who could not simply hop on the Metro to get to a doctor.

Unfortunately, and to the general displeasure of all involved, Tovar's official launch had been delayed. The medical machinery had whizzed away as expected but navigating the local roads and mountain weather had proved less straightforward. The road became slipperier than an eel's back when insulted by a few drops of water, and the trails leading to the surrounding villages were more works of fiction than of fact. Tovar's eight driving wheels, weather prediction and all the rest were simply not coping.

On the first attempt on the first day the mid-afternoon rainstorm arrived with Teutonic punctuality. And in what would become a frustrating series of such situations, Tovar squeaked to a stop at the "curva de burro muerto" to the east of the town and decided that it would go no further. It had been years since anyone with a vehicle had got stuck at burro muerto, but Tovar was not anyone with a vehicle. The friendly neighbourhood robot simply did not know to put its front wheel in the rut on the upside of the curve like the lorry drivers and motorcycle couriers did.

The mayor proposed, based on who knows what mysterious insider knowledge, that Tovar had not allowed for the low grip of the asphalt which had worn away over the years to become more symbolic than material in nature. One of the road trainers deduced instead that the rain was leading Tovar to select an incorrect apex point for its trajectory through the curve. A few malcontents from the village just above the curve insisted that the government had provided a defective tester, keeping those which worked properly for the rich people who lived at the foothills of El Avila.

By the second week the road trainers sent out by the ministry had managed to get Tovar to navigate the main road adequately, and a few of the side roads, even in the rain. The head trainer puffed his chest, bounced on his toes and nodded with satisfaction when his instrument finally read out "fully trained and verified". And for a few weeks everything seemed fine and the town's backlog of medical tests began to drop.

Colonia Tovar was even becoming something of a healthcare bright spot. Journalists meandered in to see the miracle of modern medicine in action with their own eyes. Nearby towns sent formal letters of complaint to the health ministry lamenting their neglected condition, demanding their own testers. The mayor began planning to leverage Tovar as an incentive to lure doctors away from the capital with promises of country living and unrivalled diagnostic support. A unending wave of positive psychic energy seemed to have washed over the town. That is, until the seasonal winds arrived.

Of course the road trainers did not want to return. They protested vigorously that Tovar was fully road trained. They cited the long list of other testers which needed training and the added costs to the ministry and how the ministry had left them short staffed for more than a year. They dug in their heels that whatever afflicted the tester must have something to do with how the locals were using it. Maybe children chided it, or maybe a farmer placed tree branches in its way, or maybe someone had screwed around with the sensors. Whatever it was, it had nothing to do with the road trainers.

But the complaints and accusations only continued to roll in, steady as the Orinoco. "Colonia Tovar's tester is proof of government corruption" was one journalist's contribution. And the town's information page protested "A local service which can't reach the locals". And at one point Tovar's front window ended up adorned with the word "cobarde" painted in big black letters. Eventually a troupe of town leaders travelled down to the capital to protest directly to the health minister, blockading her office door for a whole afternoon.

Eventually, grudgingly, almost spitefully, the road trainers returned. On calm days Tovar's eight wheels seemed quite at home and happy on the mountain roads, driving around potholes, gripping to ruts, and avoiding animals and tree branches. On calm days the Tovar Tester seemed a case study in how well autonomous vehicle technology can work. But when the mountain's leaves stirred under the assault of a tropical wind, it was another matter. When the crackling, crinkling and rustling started and the parrot's squawked and the monkeys howled, Tovar stopped. Completely. And would not go forward or backward no matter what verbal or manual inputs were hammered in

Soon a further month had passed and the backlog of testers to train had become too long to think about. Emotional outbursts had become common as had long unpleasant evenings in the tavernas where road trainers and locals kept a healthy distance. Occasional fights broke out. And townspeople resumed their previous ritual of long bumpy trips down to the capital to get tested. And the complaints continued to grow in number and ferocity.

It was a Tuesday, around 11.15 in the morning, when Tovar suddenly leapt into action and rushed off towards town. Jaws dropped, eyes bulged and heads swung to follow the escaping form. The road trainers had been in their vehicle checking their data for the thousandth time when Tovar simply took off of its own volition. They had not bothered to shut Tovar's systems down, assuming that the robot was stuck and would remain stuck. Instead, they were stunned to see it go missing. By the time they got out of their vehicle there was little left in the road beyond a few leaves and the acoustical accompaniment of bothered birds.

After a multitude of assessments and adjustments, experiments and improvements, a chance event had finally lent its hand to solving the mystery of the random robot. Once in town the inspection of Tovar's log files confirmed a battery shutdown. A battery shutdown brought on by the 1,800 metres and the easterly wind. A temperature regulator on the side exposed to the wind had let the team down. One of Tovar's few systems which was not literally coated in sensors and actuators. Probably the last thing the road trainers would have considered. A drop in the ocean of Tovar's complexities. The town had proven too much for the city-bred contraption.

"Blast that bunch of jerks", mumbled the head trainer once the fault had been diagnosed. "Don't they know that testers have to work in mountains and not just in warm little garages downtown". He went on to repeat the phrase. Then again, deafeningly and defiantly enough to ensure that all the townspeople who had gathered around would hear. "Blasted engineers"! he was still cursing as the crowd began to disperse.

Dustoff Danny

More than the low-pitch thuds and occasional nearby staccato it was the eyes which were becoming unnerving as the hours passed. Why ohhh why did they put such big eyes on the thing? It might be a life saver but does it have to look straight at you? All the time? Bad enough to be immobile, waiting for a chance, without having the thing staring. Such thoughts accompanied him through the night of what the journalists would later call the "minor skirmish".

A minor skirmish could of course lead to a major death. And that was exactly the risk he was now facing as he lay behind the muddy mound waiting to be retrieved. Waiting with Dustoff Danny next to him. Immobile, silent, staring. Not a clink or clank of a track, not a peep from a servo, only staring.

Dustoff Danny was a medical evacuation robot. A small tracked wonder in green and grey with room for a soldier on each side protected by armour plating. A small set of optical, thermal and laser sensors helped it make sense of its surroundings. It could not kill anyone unless it ran over them by mistake, but it was even more cherished for it. If someone was hit or down with infection or too ill with some godforsaken disease to proceed, Danny would be dispatched. And Danny very rarely failed to arrive. And they loved Danny for it.

During the current so-called minor skirmish Danny had been a very busy robot. Already on its third dustoff. And dustoff number three was proving more involved than the earlier ones of the night. Longer, more exposed and more likely to attract fire. Given the risk of damage, or worse, it's a good thing that friendly neighbourhood robots don't feel fear. Or do they?

Danny snaked around the rise, hid among a group of trees, then sprinted across an open patch of sandy terrain. A minute or two of motion at most, but enough to be spotted. Flora is rarely angular and mice don't knock down trees, thus it took hits from small arms which scratched it up a bit, but, fortunately, damaged nothing serious. Upon reaching him Danny turned ninety degrees sideways to provide cover, then stopped dead in its tracks and waited. Waited staring. Waited staring at him to see when he would initiate the evacuation sequence in the regulation way.

Dustoff Danny and his bulletproof brethren couldn't talk or signal with light without announcing their position and bringing down a shower of shells and rockets. To protect the soldiers the robots were silent, darkened, thermally insulated and stealthy at radar frequencies. A wounded soldier needed even more protection than one who was still in the fight. It was thus left to the eyes to signal the robot's intentions and to provide confirmation of having understood whatever it was meant to understand. If perhaps not the windows to the soul, the eyes were nevertheless the windows to survival.

As the soldier waited for the right moment to attempt the evacuation the dissonant soundtrack from his long humid training sessions at jungle camp percolated through his mind. The barking voice of the barking mad master sergeant was repeating that it was the face. The face of the robot would let you know what was happening. Always watch the face and never ignore when it changes. The face was the key to survival. And it applied both ways. Danny would not attempt to evacuate until the command was mouthed, eye code blinked or triggered by panic from causes unknown to the robot.

But as the hours passed those life-saving eyes became unnerving. No movement, no expression, no soul. Just eyes. A human medic might have tried to be as quiet as a mouse but the sod would be breathing. And would bite a lip from time to time as a shell came too close for comfort. Or maybe shake or cry or whisper a prayer. Instead, Danny just sat there. Just sat with expressionless eyes.

Finally the moment seemed opportune. The crashing and thumping of shells had moved westwards, dropping in pitch and fading in intensity. The danger seemed more distant. More remote. Less existential. Either his unit had moved onwards or the insurgents had decided to withdraw. Either way, maybe, probably, nobody was scanning in their direction and nobody would be the wiser if they moved out. Good fortune seemed to be smiling on Dustoff Danny and it's conscript cargo.

Faces met and orders were imparted. A sequence of eye blinks told Danny that it was time to jump into action. Electric motors hummed, an armoured hatch went vertical and the octopus-like assortment of lanyards and lifting lines deployed. From somewhere deep in Danny's bowels there were clicking sounds coming from air defence systems and a faint smell of burnt rubber. Sensors on, chaff in place, drones ready to fly. All that was left to do was to endure the struggle in, close the hatch and hope for the best.

Within seconds the two were off at speed and changing course at irregular intervals, zig-zagging in defensive manoeuvre. Suspension stiffnesses were adapting and buffer stops were adjusting to keep the cargo as comfortable as possible. A bump here, a jolt there, but a reasonable ride nevertheless which minimised the medical maladies. And other than Danny's clinking and clanking there was silence. Welcome, merciful, life-preserving silence.

Danny retraced its way back across the open patch, back through the tree line and around the rise. Mercifully, radar had noted no aerial threats and thus no need for chaff or drones. Entering the medical compound it headed for the surgical tent positioned against the backside of the hill, and doctors rushed forward for the handover. Soon the support squad would be chalking another stick figure on Danny's side plating to indicate the further life saved.

Medical Transfer Shuttle 42

The six feet took turns frictioning the floor and the synchronised gasps were audible as the clock hand snapped past the hour. Mother, father and son were in a semi-panic as they waited for the thingamabob to arrive. But there was still nothing out front.

Grandad, the family icon and clan leader, had complained of pains earlier in the evening but at first nobody had taken too much notice. He always had pains. Lots of them. Everywhere. All the time. He was in a terrible condition for someone who the doctors all claimed was in top health. So much so that the term "grandad pain" had become family shorthand for something which did not exist, a non-entity, a ghost's ghost.

But as the laments increased in frequency and intensity something had to be done, even if only for the sake of scaling the moral high ground. Ignore the fact that this had happened before and turned out to be nothing. Ignore the fact that he was in excellent colour and was moving about better than a 400 metre sprinter. Ignore the fact that it was almost midnight. Place the call.

The local health avatar quickly materialised and took control of the room. Blood pressures were measured, oxygen levels were investigated and most of the fidgety little plastic packets from the home medical kit were opened and operated. From somewhere in the bowels of the central medical database a diagnosis was conjured up which suggested a situation meriting monitoring but not construing concern. A category 4 in the new regulatory terminology. Move to clinic, take supplementary samples and monitor. A few hours of effort to provide piece of mind. And, as required by a category 4, a shuttle was issued to transport the patient comfortably to clinic.

Medical Transfer Shuttle 42 finally arrived at a couple of minutes past midnight. Engraved with huge red crosses and billboard-sized warnings about function and road rights, it could hardly be missed. As befits a category 4 shuttle it had large doors, smooth interior and medications which supplemented what everybody had at home, by law. Water, oxygen and a few drugs were also on hand as was a robotic arm to place biosensors or provide injections. At the front a large screen provided a live feed to the human transfer nurse who kept an eye and answered any questions.

Grandad was in, buckled up and off within minutes. Father and mother glanced at each other and forced a semi-smile as they bolted their home tight and activated their vehicle. A prognostication of 15 minutes' journey time was provided on all screens and by all avatars. All seemed in order as MTS42 banked down the barren boulevards, rushing Grandad to clinic.

Just past the stone bridge Grandad's coughing diminished and his pains appeared to abate. And his questions and renewed ramblings stimulated a general outbreak of relief as the several faces which were nearly touching the several screens at the several locations permitted themselves a upwards curl of the lip. The transfer nurse's methodical mind even jumped ahead of itself to begin moving a hand towards the selector switch to cancel the mission and return the patient back to where he had come from.

But as the hand nearly reached the apex of its mid-flight trajectory to the selector, Grandad slumped in his harness and appeared to stop breathing. Colour drained, motion stopped and monitors flatlined. The eyes of the several faces at the several locations exploded dangerously open, hands shook and backs straightened. Near panic ensued.

The transfer nurse struck the enormous red category 1 button activating the shuttle's emergency lights and sirens and accelerating the robot to the maximum speed allowed by law. A human driver might have pushed well past the speedometer's safe zone, but given the brutal nature of robot impacts from control system failures the law had set the robot speed limits low. Ignoring laws, rules, roads and personal safety, however, the wave of human family members shot their way towards the hospital where Grandad was now directed.

As dictated by the category 1 rules the emergency oxygen supply dropped from the ceiling and the robotic arm fixed the coupler to Grandad's face. MTS42 then applied several biosensors and initiated an intubation for stimulants. Electrical motors whirled, screens filled with red-lettered text and high-pitched warning sounds emanated from corners of the cabin. A team of emergency doctors rapidly locked into the video feeds, sensor feeds and metabolic algorithms. Grandad was now the spectacle at the centre of a virtual circle of people and equipment.

Within exactly 11.05 minutes from leaving home MTS42 was at the hospital accident and emergency ramp. Staff descended, orders were imparted, and hands and arms went through the motions. Pulses were checked and rechecked, injections were shot in here and there, pumps were prepared and specialist staff from the various corners of the facility were thrown into the fight. Grandad had suffered cardiac arrest. Distant light years from the original category 4 classification.

And he was now deceased.

MTS42 had been equipped with an abundance of drugs and devices. And the virtual circle of staff who had worked to save Grandad's life in transit were, mostly, the same individuals who later descended upon the accident and emergency ramp. But cameras and actuators are not eyes and hands. Medical Transfer Shuttle 42 could not perform heart massage or insert the ventricular assist pump.

On the hospital emergency ramp a small group of onlookers had gathered despite the early hour of the morning. "Why was this man in a category 4 shuttle"? screamed one visibly shaken doctor. Another instead asked, "Why does this type of shuttle not have a cardiac emergency unit"? in a somewhat defeated voice. Heads slumped and a tear or two shimmered. And one of the administrators who had been watching the scene evolve from the nearby sidewalk stepped into the building, linked to the hospital's legal office, and requested an urgent meeting despite the early hour of the morning.

Happy Valley Health

It was an expensive service, almost a week's salary by his calculation, but what alternative was there? His ship had not so much sailed as sunk. Things were headed downhill and were picking up speed dangerously. And there was simply no way to relax in the city and not a quiet corner to be found, regardless of what the tourism board projected on walls. The city was not a place where a person could think, never mind work through anything.

Wei had suggested that he be picked up at the corner of Princess Margaret Road and Argyll Street such that the route would be straight south on to the island. Avoid the little roads and the parked delivery vans had always been his motto. Avoid getting stuck in rush-hour traffic. Don't let the city outsmart you.

At the appointed time, God knows how given the afternoon confusion, the pod arrived. It slowed to a standstill at the sharp edge of the street, risking a bad-mannered bump from behind. It squeezed close enough for Wei to get in and within seconds it was off again, diving into the river of Hong Kong traffic, navigating the centimetres left by the pods, shuttles and the odd human-driven vehicle.

It was the first time that Wei had given this particular service a go, but he was quickly taking a liking. Cloth-covered surfaces, soft seats, wonderful woods, a drinks dispenser which could put a convenience store to shame and soft reddish orange carpets. Nothing like the plasticised environs and sanitised smells which made a person queasy when visiting a local mental health professional.

And the lighting was spread over the inner surfaces like jam on bread, a continually changing colourscape of fields, forms, bubbles and blotches. Was that a swarm of fish at the front? Or perhaps a group of clouds? Or two dragons fighting? At times the shapes moved left to right or top to bottom or diagonally, as if words of ancient text. At other times they circled or hovered like bad omens. Now the front corner was turning slightly yellow, honey-ish in fact, then it and the rest of the cabin somehow became light brown.

And when amorphous flower-like things emerged from the mist they were always accompanied by olfactory outbursts. Aromas permeated and penetrated the cabin, boring deep into the psyche. Squiggling and wiggling their way inside. Perhaps unsurprising in the city of the "fragrant harbour", the biting bouquets and spicy scents were exotic enough to merit outright rejection from a pragmatist's pragmatist like Wei. But instead on this occasion they mingled, mixed and merged with Wei's less than optimistic mood.

The pod was now in Wan Chai and it's electric motors were dropping in pitch from the gradient of the climb. Though what seemed like a grandfather clock's worth of time had passed since he had entered the pod, it had not yet spoken. "Welcome to Happy Valley Health's relaxation drive, is there anything about my driving or the cabin or the music which you would like me to adjust"? it finally muttered in a voice something between a female secretary and a male chauffeur. "Is there any food or drink which I can provide you Mr. Wei"? followed in the same difficult to describe voice, but in noticeably more neighbourly intonation. All pods checked needs before engaging in any form of conversation. It was a legal requirement. A city ordinance in fact.

As they continued upwards the world turned from grey to green and the oncoming traffic melted away in obedient respect of the mission. The interior surfaces transformed into windows which at certain curves surveyed the sea or pictured the peak. Sounds now penetrated into the cabin, bringing with them the rhythms of the road and the murmurs of the mountain. Like an astronaut entering orbit he now hovered, weightless, in sensory space. Floating free from abusive anxieties.

"What brings you to the island"? asked the pod as it started to deploy its interaction schema.

For a time Wei left the question unanswered. His response was slow to arrive, hesitant, unsure of whether there was more. Like everyone else he had become accustomed to verbal exchanges with pods, robotic roadside vendors and humanity's many other artificial assistants, but this was the first time that he was talking, really talking, to an automation.

"I had a bad week at work and I lost an export contract … and it got me into an argument with my boss and then also with a neighbour. I can't quite seem to reboot", he mumbled in a tone lower than the ground beneath. "Thought a trip out to the island with someone who listens might help".

Happy Valley Health prided itself in fielding the city's best fleet of mental health support pods. Other providers had ringfenced a few of the trendiest scenic sites and culinary cookeries, but Happy Valley Health was second to none in terms of its conversations and psychological support. No other service could help a person reflect upon difficulties and find healthcare specialists as well as Happy Valley Health. And Wei's pod, in particular, had already achieved a remarkable record. It's on-board algorithms estimated that over the last month alone it had cheered its passengers by 3.2 points on average on the ten-point Happy Valley Happiness Index, and had made 11 referrals to mental health experts. If a Happy Valley Health pod could not help, then nothing could.

By the time the pod reached Stanley, revealing, even intimate, conversations were under way. "How do you feel about that person"? might ask the pod. And "maybe I shouldn't be telling you this but"… might respond Wei. As with any conversation, this one swung back and forth like a branch in a breeze, stopping at times, speeding up instead at others, maintaining its own rhythm and producing its own surprises. The pod seemed unusually good at waiting the right amount of time for answers, not interrupting, and reworking or repairing the conversation in a nearly human manner. A good talker. And, above all, a good listener.

Wei could almost see the stress oozing out through phantom pores. It did not occur to him that the pod was correlating causes and effects, analysing constraints and searching for mitigations. The science escaped him. And perhaps he did not care. All that mattered was that the ride was proving relaxing and that the pod's suggestions seemed useful. And if the pod thought that one of the specialists in Central District was qualified to provide support against his occasional bouts of depression, then why not try? How could any human be better informed than Happy Valley Health about the local mental health scene?

In what seemed atomic clock time the pod was back at Princess and Argyll and it was almost dark. Where had the afternoon gone? When you lose an export contract the problems seem to go on for months, but when you do something pleasant it is over before you start. Wei shielded his eyes from a knifelike "new neon" sign as he followed the ritual of the pod deactivating its systems and locking them into place. Like a peacock, it smoothly manoeuvred various bits shut, locking them with a low-pitched hum. Then, just before it launched itself back into the traffic stream, it spoke just loud enough for Wei to hear. "Happy Valleys Health is proud to have been of service".

References

artefactgroup 2023, AIM healthcare in the age of AI, artefactgroup, Seattle, Washington, USA, https://www.artefactgroup.com/case-studies/aim.

Bannerman, L. 2023, A&E on wheels brings the hospital to you – the Times Health Commission examines a revolutionary model of A&E treatment in East London, *The Times*, https://www.thetimes.co.uk/article/times-health-commission-aande-on-wheels-skoda-vdbn5pc6k.

Blanco, S. 2020, NAVYA's autonomous vehicles are transporting COVID-19 tests, *Car and Driver* magazine, April, https://www.caranddriver.com/news/a32051316/navya-autonomous-vehicles-coronavirus-tests.

Cenizo, S. 2021, VW's autonomous ambulance will help save more lives, *CarBuzz* magazine, October, https://carbuzz.com/news/vws-autonomous-ambulance-will-help-save-more-lives.

Elayan, H., Aloqaily, M., Salameh, H.B. and Guizani, M. 2021, Intelligent cooperative health emergency response system in autonomous vehicles, in IEEE 46th Conference on Local Computer Networks (LCN), Edmonton, Canada, 4–7 October, pp. 293–298.

Huang, Y., Yuan, L., Tang, J. and Liu, S. 2021, Application of autonomous driving technologies in the presence of COVID-19: to reduce occupational exposure and potential nosocomial infections of care workers, during transportation in Emergency Medical Service (EMS), in IEEE International Conference on Consumer Electronics-Asia (ICCE-Asia), Penghu, Taiwan, China, 1–3 November, pp. 1–3.

Khalid, M., Awais, M., Singh, N., Khan, S., Raza, M., Malik, Q.B. and Imran, M. 2021, Autonomous transportation in emergency healthcare services: framework, challenges, and future work, *IEEE Internet of Things Magazine*, Vol. 4, No. 1, pp. 28–33.

Lim, Y., Giacomin, J. and Nickpour, F. 2021, What is psychosocially inclusive design? A definition with constructs, *The Design Journal*, Vol. 24, No. 1, pp. 5–28.

Linder, C. 2020, An unlikely coronavirus hero? Self-driving cars – when humans fall ill, it helps to have some robotic help, *Popular Mechanics* magazine, March, https://www.popularmechanics.com/technology/infrastructure/a31401964/self-driving-cars-coronavirus.

Liu, S., Huang, Y. and Shi, L. 2022, Autonomous mobile clinics: empowering affordable anywhere anytime healthcare access, arXiv preprint arXiv:2204.04841, https://doi.org/10.48550/arXiv.2204.04841.

Winter, S.R., Keebler, J.R., Rice, S., Mehta, R. and Baugh, B.S. 2018, Patient perceptions on the use of driverless ambulances: an affective perspective, *Transportation Research Part F: Traffic Psychology and Behaviour*, Vol. 58, pp. 431–441.

Chapter 9

Providing Entertainment

Background

As noted earlier in this book the safe transporting of people is the focus of most of the current autonomous road vehicle research. However, once safe and reliable autonomous road vehicle platforms are achieved, the creativity and curiosity turn naturally to the question of what else the friendly neighbourhood robots might do for people beyond transporting them. And one human activity which might benefit from further automation is that of entertainment.

From sightseeing buses to Nashville street parties (Rojas 2021) many human-driven road vehicles are used for purposes of leisure, tourism or entertainment. So called "transportainment" (Schukert and Müller 2006; Beyer 2016) vehicles blend the function of transport with that of entertainment. But while the human-driven minibuses, tourmobiles and other such vehicles provide the mobility, the didactic or entertainment elements of the experiences are usually currently provided by human professionals. Not by the vehicle itself.

In recent years, however, there have been suggestions that autonomous road vehicles can provide entertainment opportunities. The elimination of the dedicated human driver will lower the expense and increase the privacy of a range of services. And with improved automation the autonomous road vehicles may soon provide the leisure or entertainment functions in addition to providing the more traditional point-to-point mobility.

Future friendly neighbourhood robots can be designed to be autonomous also in the non-driving aspects of the leisure, tourism or entertainment experience. The capable computerised companions can perform the multiple tasks of a single human, and perhaps even the multiple tasks of multiple humans. Since improvements in automation and in artificial intelligence are transforming road vehicles from the realm of the mechanical to the realm of the social, multiple interactions which have little or nothing to do with physical transport will soon be possible.

DOI: 10.4324/9781032724232-9

One line of reasoning involves what can be called "leisure robots" or, if even more highly specialised, "gaming robots". The satisfying of the human desire for leisure and entertainment has been ongoing in all societies all through history. And as examples from the *Star Trek* holodeck to the *Westworld* robots suggest, it has also been a prominent feature of works of science fiction (see, for example, Shedroff and Noessel 2012). The vast assortment of systems found among the pages of the science fiction literature suggests a likely demand for such services in the real world. Given the many leisure and entertainment concepts found in science fiction it would be surprising if at least some of them did not become manifest in science fact.

"Leisure robots" are minimally modified machines which provide modes of human entertainment within a comfortable or even luxurious cabin setting. Such friendly neighbourhood robots typically include comfortable and reconfigurable seating systems, multiple video screens, sophisticated sound systems and significant on-board computational power and network connectivity for communications and gaming. Such machines prioritise comfort, relaxation, enjoyment and entertainment, often above and beyond the transport task.

Typical of the many current proposals for leisure robots is the Cadillac SocialSpace concept car (Cadillac Corporation 2023) for which the company claims "Biometric sensors placed throughout the cabin interpret passengers' vital signals and adjust interior temperature, humidity, lighting, ambient noise and even aromatics to match their moods" and "…acts as a living room on wheels, with eye-to-eye seating for up to six guests and lavish amenities that allow travellers to recharge and reconnect with one another enroute". Such developments suggest an increasing integration of automotive, smart-home and smart-office technologies to provide immersive and interactive experiences.

"Gaming robots" are instead maximally modified offerings which provide state-of-the-art digital connectivity and advanced gaming capabilities. Work is currently under way to incorporate existing gaming systems within autonomous road vehicle platforms, often taking advantage of the increased cabin design freedoms to achieve larger or more encapsulating screens and projectors than possible in home environments. And research is also under way (see, for example, Lakier et al. 2019) to develop new games designed specifically for use in autonomous road vehicles, which take full advantage of the peculiarities of the road travel itself.

While few fully arcade-like concepts cars have emerged to date, developments in on-board systems and in gaming technology point to where the future friendly neighbourhood robots may be heading. For example, BMW and AirConsole have collaborated to integrate the AirConsole gaming platform and the BMW curved display in human-driven vehicles. The integrated system provides both single and multiplayer games which can use a smartphone as the controller (Brook-Jones 2022). And Nvidia has announced that its GeForce Now cloud gaming service will be available in human-driven vehicles from motor manufacturers including BYD, Hyundai and Polestar (Takahashi 2023). Such developments suggest the increasing integration of the individual automotive and gaming technologies to provide more complex and more immersive experiences.

Another line of reasoning involves what can be called "autonomous tourmobiles" or "tourism robots". In recent years the possible benefits of automating not just the travel itself but also the conversational, didactic and educational activities of tours has been discussed. Automation should soon permit the vehicles to perform those activities without the need for the traditional human tour guide. In fact, one organisation which operates in the leisure, hospitality and tourism sector, Autoura, has already developed an AI companion for such purposes. Available in the Apple and Google ecosystems the Sightseeing Autonomous Hospitality Robot by Autoura (SAHRA) provides programmable voice and visuals which can be deployed to provide information and answer questions in human-driven and autonomous road vehicles.

The concept of autonomous tourmobile has increasingly entered the friendly neighbourhood robot discourse as digital mapping, local area sensing and on-board computational power have improved. Cohen and Hopkins (2019) have, for example, suggested that "All types of vehicle transport involved in urban tourism will be affected by the potential transition to automation. This ranges from airport shuttles and transfers, through city taxis, car hire and vehicle-based guided urban sightseeing". And several business cases have been outlined (see, for example, Bainbridge 2018) for such offerings.

A further line of reasoning in relation to providing entertainment involves what can be called "party robots". Cohen and Hopkins (2019) have, for example, suggested that "Restaurants may find themselves in competition with connected autonomous road vehicles that become moving restaurants, or combine urban sightseeing with dining – as exist today with dinner cruises". And noted that "Stag and hen dos may become spread out, as opposed to concentrated in particular bar districts, and reliant on connected autonomous road vehicles to move drunken revellers across greater distances between drinks in the urban night, perhaps even crossing multiple cities". And warned that "... prostitution, and sex more generally, in moving connected autonomous road vehicles, becomes a growing phenomenon. For instance, 'hotels-by-the-hour' are likely to be replaced by connected autonomous road vehicles, and this will have implications for urban tourism, as sex plays a central role in many tourism experiences".

Party robots are maximally modified offerings which provide state-of-the-art venues for specific forms of human socialising. The typical premise is that an affordable mobile party can act as a multiplier of the enjoyment derived from similar activities at a fixed location. Motion and mobility providing additional functional, sensory and cognitive dimensions to the experience. The physical characteristics of the robot's interior are designed with the specific form of socialising in mind. And as with other forms of friendly neighbourhood robot the automation is used to reduce the cost, increase the privacy and support the intended human interactions.

While there seem to be some human-driven road vehicle precedents for the new services, few autonomous party robots have so far been experimented. It is thus not yet clear how effective they will be in their intended role or whether the multiplying effects of autonomy will be confirmed in practice. However, it is not difficult to imagine a target metaphor of a 1980s video arcade on wheels or of a small exclusive nightclub. Privacy should prove somewhat straightforward to achieve and costs may prove competitive.

And a final line of reasoning in relation to providing entertainment involves what can be called "companion robots". Companion robots are maximally modified offerings which provide state-of-the-art conversational, emotional and social capabilities to interact with people in socially relevant ways. These capable computerised companions will be optimised more in terms of their linguistic, emotional and social interaction capabilities than their road holding or their driving style. The objective will be squarely within the realm of the social, and in some cases may have little or nothing to do with physical transport.

Recent years have seen the growing popularity of pet robots such as Aibo (Fujita 2001) which have become companions to thousands of people. More so than robots in factories or hospitals, pet robots are designed and produced specifically for the purpose of providing entertainment, companionship and even friendship. As these small robots have repeatedly revealed in practice, companionship and friendship do seem possible between robots and humans.

If such relationships were to prove desirable also in the case of the larger and more capable friendly neighbourhood robots, then a specific genre of vehicle will emerge. The use of sensors which can estimate human emotions will prove vital as will the ability to respond to people in anthropomorphically emotional manners. The increasingly popular profession of conversation design will also prove key to success. And an additional set of technical, philosophical, ethical and legal questions will be raised for the designers to deal with, and for society more generally.

While few fully companion-like concept cars have emerged to date, developments in sensors and on-board emotion systems point to where the future friendly neighbourhood robots may be heading. For example, Kia has adopted some of the needed technologies with its Real-Time Emotion Adaptive Driving (READ) system which monitors passengers to adjust driving and cabin settings (Billington 2018). And Toyota exhibited it's Concept-i which has built-in artificial intelligence which is claimed to learn the driver's needs and habits (Hawkins 2017). Concept-I learns locations, driving patterns and emotions. Then uses emotionally appropriate lighting and sound to inform the driver about matters such as the vehicles' settings and road hazards (Morby 2017).

Also pointing the way to what the future companion robots might be like are the current generation of artificial intelligence-based social chatbots. Several social chatbots can already be downloaded from the App Store or from Google Play to be used across a range of digital devices for purposes of social interaction or even companionship. More so than the more intermediary offerings such as Alexa and Siri, these systems are specifically optimised for linguistic, emotional and social interactions.

The Replika chatbot, for example, requests that its users answer a set of predefined questions, then uses the information in its written and spoken conversations (Murphy and Templin 2017; Luka Inc. 2023). Once Replika's responses become systematic and mature, it starts to become a companion which is always available and always willing to discuss any matter. Providing something of a therapeutic vent for expressing concerns, fears, interests and aspirations. Along similar lines the Woebot mental welfare chatbot adopts principles from cognitive behavioural therapy (CBT) to help manage distressing thoughts and feelings (Woebot Health 2023). After inputting various required items of information, the Woebot acts as a mental wellbeing-aware companion which provides a therapeutic vent.

Leisure robots, gaming robots, autonomous tourmobiles, tourism robots, party robots and companion robots are all forms of friendly neighbourhood robot which seem likely to appear on roads in the coming years. Such developments are likely consequences of the increasing demand for cost effective and flexible opportunities for leisure, entertainment and social interaction.

What follows on the remaining pages of this chapter are a set of four science fiction prototypes, short immersive vignettes which explore the nature and implications of the future friendly neighbourhood robots. Each involves physical, psychological and social interactions with the future friendly neighbourhood robot and several allude to anticipated ethical conundrums.

As discussed earlier in this book the science fiction prototypes are linguistically based and avoid an excessive focus on the robot itself, so as to leave ample room for imagination and speculation. It is the author's hope that they can provide a focal point for discussion of autonomous road vehicle design.

Privacy Pod

The call reached me smack dab in the middle of extracting our local celebrity darling from her self-inflicted predicament, but a pinch of time was more than enough to realise that Adult Encounters needed my services. Vacillating vocals suggested that they were plunged in problems and sinking swiftly.

Prostitution, sex work and adult entertainment are among the names which have been used. History is speckled with money-making en-counters in villages, towns and cities across the world. These days the encounters are more complicated than they once were, involving feeds, passwords and electronic transactions, but the human stories are much the same. As are the human problems.

And one answer to one problem has been the privacy pod. It all started with a newspaper article which claimed that it was the getting here and the getting there which exposed people. When not in the e-privacy of one's own home, the travel to and from the encounters prompted the greatest perils. It exposed people to aggressions, scandals or just the simple embarrassment of their snoopy neighbours seeing them when walking by. For such maladies the privacy pods were the ideal remedy.

They were originally launched as multipurpose machines for anything requiring a coat of confidentiality. Weekend romantic getaways, drunken office parties, or anything silly enough or intimate enough to suggest avoiding the public eye. But, as always, Johnny public evolves anything which is released into the wild. And the respectable romantic robots were no exception. They morphed into the vehicular venue of choice for illicit encounters and for those providing adult entertainment. Privacy, mobility and convenience. A new trend was born. One which keeps me busy and which pays my bills.

The mishaps, e-barrassments and cyber shaming now swell my revenue stream from the dusty riverbed it once was into rapids which flood across my bank manager's screen. My skillset, a bit of cyber-privacy and a bob of cyber-sex, hits the sweet spot. In the right place at the right time, so to speak. And, astonishingly, I am still the only guy in town who can put his finger on what's going on and anticipate how the clowns will react. To my great satisfaction, this latest wave of incidents has grown my revenue stream into what now resembles a mighty Mississippi of money.

After the call from Adult Encounters my first stop was their Capp Street garage where I was briefed on the situation by their overworked and now zombie-like head technician. His recent days had obviously been spent connecting privacy pods to diagnostic devices, and his nights must have consisted of unending readings of log files and sensor values. I had hoped that he might have spotted something by now, anything, which could be used to unwind the thread back to its source.

But no electronic devices had been found which were not supposed to be there and no wires or connectors were out of place. Version 6.2 of the pod's control system was in place and operating correctly, and only the company itself and the last three clients were anywhere to be found in memory. According to every piece of information which he had managed to squeeze out of the pod the thing was configured correctly and working the way it did on the day that it arrived from the factory. Not a sign of tampering and not a hint of temperamental behaviour.

And yet, just as the journalists had written about the clients of the other pods, embarrassing images had begun to appear. An offence made worse by this Adult Entertainment pod having also performed the cardinal sin of being seen driving through an ill-reputed neighbourhood where it should never have been. Everything suggested a case of the newest game in town, sextortion. People sabotaging privacy pods for profit. Threats of reputational ruin. The technologically savvy culmination of a long history of sex smeared scams.

A shipload of colourful materials were already rumbling their way around the rumour rooms. Enough fabrication of insinuations and innuendo to risk bringing down the electrical grid. And the influencer who was threatening Adult Encounters via her legal team was currently on the offensive, creating a tricky situation which the company was finding difficult to extricate itself from.

And of course the knee-jerk response had been the usual "blame the pod". The now prevailing practice of getting out of hot water by putting a pod in it. Fingers were pointed at the manufacturer while toes turned towards the imperious influencer. Somehow the pod must have misbehaved or been manipulated, that was the first line of defence. The updated version of the usual picking on the little guy.

It would be a nightmare for the company if the real culprit were not found soon. Adult Encounters was a franchise, part of a larger operation with privacy pods all across the country. Lawsuits would emerge east and west and everywhere in between. They risked being the first defendants of the first case of cybersecurity failure involving public defamation. A whole new realm of litigation and a whole new branch of bad press. Enough to keep the headache tablets popping late into the night. And, of course, more than enough to justify my reasonable rates.

My second stop was over in Oakland to see Sohail from the manufacturer's coding team. Like Adult Entertainment's own head technician, the wild-eyed coder had probably not seen much daylight of late. His steadfast conviction was that an adult entertainer had carried a personal device on-board and had hijacked the pod's transmissions. The device had probably needed only a camera and a tracker and the right code to send the pod off on its self-incriminating saunter.

But the data simply did not match the speculation. There was nothing odd with the data types and nothing had changed with the data transmission rates. And, as he should have known if he had actually been earning his pay, the pod's bodywork contained a metallic cage which blocked the direct electromagnetic communications of the passengers. Expensive, but needed, given the nature of the vehicle. God knows how many times I had heard such silly suggestions from code junkies over the years. And only God knows how they hang on to their jobs.

My third stop was back in the Mission District to interrogate Julio from Adult Entertainment, the pod's designated engineer. Part master and part servant, he was the one person who would really know the pod. His responses to my questions would not be based on physics, or engineering, or economics or some other pie-in-the-sky activity, but the result instead of grease and sweat and doing whatever one does when living with a thing from day to day.

And his specific slant on the matter was that some bug deep inside the guidance system could have triggered the sequence of events, misleading the pod's other systems like a drunken conductor. But, again, this angle of attack did not seem to me to match the diagnostics. Such a fault should show, somewhere, somehow. From years of dealing with puckish pods I know for sure that the timing circuits are simply too sensitive to not deviate, at least a little, when something is going on inside. A drunk can't do additions or subtractions properly no matter how hard he tries.

But the hour or two of Julioing had not been a waste of time. Something he said about his most recent safety servicing led me to check, I'm not really sure why, the video feeds which monitor the occupants in those few seconds when a road hazard happens along. Why was the incendiary influencer who was at the heart of this pod's many maladies always out of view? How, or better yet, why, would she be off-screen every single time? In my experience people don't cramp in corners no matter what weird and wonderful things they are getting up to. The Julio-stimulated revelation did not of course tell me how they did it, but it did suggest that they were up to something which had nothing to do with Adult Entertainment's service.

Maybe, just maybe, I might be about to earn my very reasonable rates.

Roman

Acqui-Terme was the next stop, a sleepy cluster of 25,000 souls nestled in the Apennine foothills between Liguria and Piemonte. Now a small provincial town, in Roman times it was a large and important city where legions stopped to enjoy the thermal baths and dignitaries drank the waters before crossing the Alps. A fine first-century thermal bath is still to be seen today and the remains of the aqueduct are considered to be among the finest in the region.

This day the twenty-four torpid tourists had been awaken at first light and hunger pangs were now stimulating conversations more than ancient history. Torn wrappers, empty snack bags, and a rock garden of bottles and cans now cluttered the confines. The tummy rumblings sometimes synchronised with the road wheels and many of the travellers began to fidget to the fluctuations. More than one eyebrow was raised distinctly upwards like the sun which was just now visible on the horizon at the front right.

"Roman, which town is next"? asked the young chap who said he hailed from Normandy.

"Acqui-Terme is next, Mr. Laurent", was the reply uttered in a somewhat newsreader's voice, "it's approximately 50 kilometres from Genova and approximately 80 kilometres from Torino".

"Ok, ok, ok, Roman, but are we are stopping there to eat? And is the food any good? I am hungrier than a hungry thing".

"We will be stopping at the Trattoria Della Piazza Bollente where they offer local cuisine and selected ancient recipes", was the response in a tone which had now shifted from newsreader to testimonial influencer.

"Yes, yes, yes, that's all wonderful, but is the food any good for heaven's sake"? prodded the famished Frenchman.

"Apologies, but I am not authorised to express opinions", was the rapid reply in a now reinstated newsreader motif.

Stiffening his back and shaking his head, Laurent's voice now expanded enough to ensure that his cruising companions could hear.

"Fantastic, he knows the date, style and backstory of every little artefact in every little museum, but can't tell us if the food we are about to eat is any good".

"Given the cost of the tour why don't they have humans to help explain things"? was the final transmission from the now frustrated Frenchman, which terminated the conversation and dampened dispositions all through the cabin.

Roman was a friendly neighbourhood robot of recent manufacture but was not new in the usual sense. Its chassis was intended to reassure those who had holidayed previously on human-driven vehicles. More safety, more features, but a traditional atmosphere. Four larger than normal tyres promised grip. Large white frostable windows stretched from one side all the way over the top to the same point on the other side. The seats were large, soft and studded with all manner of gadgets to help pass the time. And unlike the old human-driven vehicles, getting in and out was a doddle thanks to a door and steps for each row of seats. No waiting in line, no pushing and no shoving. Convenient convenience.

But it was the scripting which was the tricky part. The subject of a new branch of design. Contextual awareness, in particular, was said to be where the gremlins growled and the devils did their dirty work. The child who runs to the bathroom risking a fall, the university professor who relishes rubbishing the robot's knowledge, the young couple from the other side of the world who never heard of the civilisation. A human tour guide might try to "wing it". But friendly neighbourhood robots, notoriously, don't "wing it".

And as they came off the viaduct and began cruising Corso Bagni they hit fresh turbulence.

"I don't recall Piazza Bollente being on the list of the stops", said one tourist in a heavy Swiss accent.

"I thought that the virtual tour said something about the Hotel Terme on the hillside", offered another.

And from one of the rows at the back someone jumped in with a brave "What if we want to eat at the Hotel Terme"? and someone else ranted about dietary needs and someone else about lactose intolerance and someone else asked a question so complex that only a dietary clinician could attempt an answer. The questions rumbled in like a summer storm rolling down an Alpine mountainside.

But Roman was not equipped to deal with such matters. Roman could point out the ancient wall at left, or read off a brief history of the

Roman Empire, or note a rare bird of prey flying overhead. Facts were not usually a problem. After all, that's what computer memory is for. But dealing with opinions and societal expectations were not Roman's strong points. Like humans, Roman had blind spots. And the current verbal avalanche was parked squarely in one of them.

The incongruous requests overwhelmed the logic circuits and resonated the decision algorithms up and down, left and right. Not letting them settle on any fixed probability or known heuristic. If the designers had not separated the control circuits, and provided crossed double driving, one more comment about a pizza would have driven Roman straight into one of Corso Bagni's ancient trees. And all Roman could do was to repeat "Please press the assistance button to contact a human operator" at ever increasing frequency in the manner of a frenzied fire alarm.

And contact they did. Nearly all twenty-four of them. And at nearly the same time. An avalanche of appeals rolled into the tour operator's offices in Genova, demanding the combined energies of a row of humans sitting at identical desks, in identical cubicles, under identical energy saving lamps. A check of a tour specification here, and a check of a lunch menu there, an apology for the behaviour of the tourist of row 4 to the couple in row 5, and so on. The humans scrambled to deal with the problems of the humans. And, slowly, the humans began to choke the confusion which had been conjured out of thin air by the humans.

They were now pulling into Piazza Bollente and slowing to a halt. Roman had navigated the narrow medieval street admirably, stopping for a child on a bicycle, avoiding the flower stand which had been left just a little too far into the road, and avoiding scraping against the low balconies where the road narrowed to little more than a footpath. Safe, smooth and millimetrically accurate. World-class driving which few humans could be expected to match. But, as Roman sat obediently in the blue square, no one was leaving. The humans just sat or wandered around noisily inside. It would take another 15 minutes to sort everything out and get everyone off to lunch.

Fans R Us

Music styles never really go away, not really. They are stickier than glue. They predispose populations and move the masses decades after their heyday. And K-pop is still big, very big, long since its heyday. K-pop inspired music, stories, outfits and paraphernalia of every sort flood the world. Heck, people even buy K-pop carpet cleaner and K-popping car alarms.

And a big day in the K-pop calendar had arrived. The colossal concert was being held in central Busan, not far from the station. And from sunrise fans were preparing or were already on their way. Thousands. Many thousands. And most of them were now making their way eastwards or southwards towards the city centre, some in big groups, some in little groups, and some with just a pet under arm for company.

I-Jun, Su-ho, Ha-yoon and Ha-rin had decided they would travel together. All from the same high school, they usually hung out together whenever there was a bag of time which needed filling. The trip from their village north of Geumjeongsan was characterised by countless congested changes on public transport, thus a decision was taken by unanimous agreement to hire a pod. One of the pods suggested by the concert organisers. And given that word-of-mouth had been singing its praises, they opted for a Fans R Us pod.

On the morning of the decisive day they coagulated at Ha-rin's house. It was close to everyone and equipped with all the latest audio-visual gear. I-jun was the first to arrive, kitted out with the accustomed armful of snacks. Then Ha-yoon meandered in. Then, last, as usual, Su-ho stumbled through the door with his usual air of someone unsure of where he was or why he was there. Once the head count reached four, the music started, the holograms were unleashed and the psychedelic swims in the feeds began. Soon the house was packed to capacity with physical and virtual dancers, and the energy levels had risen to match the occasion.

At the appointed time, to the second, the Fans R Us pod arrived. It pulled up in front of Ha-rin's house at the only available spot, the little blue box set aside by the city planners for the friendly neighbourhood robots. Usually the only gap along any street, the boxes were a necessary evil and steady source of tension. Complaints were continuous, arguments abounded and the occasional fist fight between residents was not unheard of. Ha-rin's uncle, who was a member of the local planning committee, even got his nose reconfigured during one such altercation.

Placing blue boxes had in fact become such a thankless task that nobody wanted to be involved with selecting the spot. For one person it was too far away to load their heavy goods. For another it was too close, leading to suffering the noises of the comings-and-goings late into the night. Given all the commotion, any visitor from outer space would be excused for thinking that the boxes were the most valuable pieces of the planet.

As anticipated the pod was small, roundish and white, with just enough seating and tables for four. The side doors seemed larger than the pod itself as they swung upwards and folded on to themselves to sit in a tidy little pile on the roof. What seemed a small staircase hummed its way out from the lower part of the pod to welcome everyone. No wheels were visible underneath, which must have been the conundrum which got the teens bending and reaching and twisting and turning as they moved around the pod and viewed it from all angles. Maybe instead of four large wheels there were eight or twelve or some other number of small wheels? Or maybe it moved over the road using something other than wheels? Like scientists analysing an alien life form, eyes opened, heads were scratched and serious sounding statements were passed around the team.

The pod did not speak. Fans R Us pod pods never spoke. It was not considered polite. Following perhaps the ancient axion that domestic help should be seen but not heard, they had been designed to respond via simple motions or short explanatory text statements shown on screens or floated holographically. Requests could be made by hammering predetermined options on the surrounding panels or by ordering the pod vocally using a small set of semantics. But despite executing the instruction, the pod remained mute.

Once inside, comfort came quickly. And with a simple push of the "off we go" button, off they went. Despite there being no need for one, Fans R Us pods all had a large red "off we go" button in the middle of the passenger compartment. Years of market research and customer testing had decisively demonstrated and scientifically settled that nothing was more satisfying for people than ceremoniously launching the pod via a great big bang on the "off we go" button. And if a Fans R Us pod is not about satisfaction, what is it about?

As they now headed southwards the friends delved deeply into the interconnection features to meet likeminded fans. In a flash they learned that several classmates were also on their way, some complaining of parents tagging along. But incandescent inter-pod clashes soon broke out over which of the music group's songs were the best. And several of the eastern incoming pods were eventually excommunicated from the tuneful tribe because of their lobbying for the wrong songs.

"Bunch of seasiders", yelled the leader of one of the northern pods.

"Mountain goats", replied an eastern pod militant.

But for the most part the occasion was wrapped in good will. And the convivial cargos of several of the Fans R Us pods had now joined up on the roadway, forming a pod platoon as they moved south. The passengers were almost physically within reach of each other as they waved through windows and exchanged avatar ambassadors. Dancing, awkwardly achieved from within the rolling and pitching platforms, soon broke out.

Ha-rin and friends ended up in something of a singing contest with the similar-sized syndicate in the pod immediately behind them. A cacophony of sounds erupted. Some pleasant, some not, but none resembling the original songs. The competition soon moved to matters of common classmates, favourite haunts and individual inclinations. Eventually, perhaps inevitably, the other pod's crew challenged them to a race.

But racing was not allowed. Delving in feeds, platooning, route deviations, stop-offs and all manner of other customer requests were handled without hesitation. But not racing. Long ago the operator had come to realise that it leads to a crush of customer complaints. Too many risks, too many liabilities and too expensive when things went wrong. No, the Fans R Us legal team would never allow it. No, non, nada, nee, nein, ni, nyet. Whatever the language, the answer was the same.

The increasingly agitated attempts at ordering the pod to race were all summarily rejected. Shouting, gesticulating, typing or banging on the pod's panels produced exactly the same reply on the main cabin screen, "Apologies but your request cannot be accommodated". Perhaps transparency and honesty might have been better served by a large red-lettered message of "Apologies but your request cannot be accommodated by the Fans R Us legal team".

But the frustration soon faltered as they were syringed through the city streets towards the gated gig. The mood had dipped, but not dropped, and the ginormous holograms on top of the venue soon shot everyone's mood into orbit. The pod would soon be dropping its four K-poppers at gate number three. And the concert would be starting in less than an hour. The serious K-popping was about to begin.

Chess Mate

It was once played using small statues on boards or table tops, but now involved geometric shapes on shiny screens. And it was once played at home or in parks, but you just didn't meet enough people that way. Thus the enthusiasts invented new ways of chancing upon their opponents, often perfect strangers, in order to play their game. Day-old technologies were harnessed to the two-thousand-year-old game to produce numberless new ways of experiencing it.

And on this particular day, May 10th, after a long pause came the dramatic declaration "black pawn to D3". A shape then moved obediently to its new spot on the screen to the accompaniment of a modulated marching melody. But the sound, which must have been louder that what the designers had intended, seemed strangely at odds with the smooth motion of the piece across the screen.

"Ohhh no you won't be catching me with that one, Charlie boy, I saw that one coming, from a mile away. Try this one on for size ... white queen to B3".

"Ok, John, maybe you did see that one coming, but let's try black queen to F6 and see if you were expecting this one as well. And, as a friend, I really have to say that you should take more time to think things through because if you keep rushing like this it will end in tears".

"Not a rush. A well-tuned response, Charlie boy. And why don't you try this one on for size, white pawn to E5".

"Ok, suit yourself. If you want to rush everything it's no skin off my back. I will now go with black queen to G6".

"White rook to E1".

"Tut, tut, tut ... such an unenlightened move. You leave me little alternative but to teach you a lesson. Black knight to E7".

"Are you using a system of some kind, Charlie boy? Scamming me? I know your kind. You pretend to not be good at anything, then run a routine on your latest sucker using moves which you memorised the night before. Dammit, Charlie boy, I know you're scamming me".

"John, chill out, think of your blood pressure. Would I do something like that? And to you of all people? You're in a really rotten mood. What happened, another run-in with the boss? Or maybe another scrap with the tax authorities? From where I'm sitting there's no avoiding the conclusion that you are the perfectly polished product of a very bad day".

"White bishop to A3".

"Black pawn to B5".

"Look, I had a perfectly normal day in my perfectly normal life and got along perfectly well with bosses, tax authorities and everyone else, dammit. You are definitely scamming me. I've played you before, but this time something's different. You must have stayed up late last night to organise a scam of some kind. I don't need to reboot my external memory to tell that something's sneaky and malicious about your moves today. But I'm going to show you what a rubbish player you are, scam or no scam. White queen to B5 ...".

"Black rook to B8".

"White queen to A4".

"Black bishop to B6".

"White knight to D2".

And on it went until a "checkmate" cracked the room in two. So loudly, in fact, as to halt all attempts at conversation and scramble any thoughts which may have been forming. The thunderclap slowly dissipated into a short dazed silence worthy of the most sombre of occasions, and seconds passed before neural rhythms reformed and words were again workable.

"You see, John, that I had nothing up my sleeve? In the end, you won the game despite all the whining, whimpering and crying like a baby. I don't recall having ever played anyone quite so annoying before, and I've definitely never been beaten by such a spoilsport. If you are going to act this way you really should have a go at a different game from chess. Chess is supposed to be calm, sophisticated, for thinking people. Not for talkers and grumblers and cry-babies like you've turned into".

"The win proves nothing, Charlie boy. You were up to something. The moves were different. I could smell it. Shifting attention on me is not going to change anything. Trying to make me look thicker than a submarine door is not going to work. I know what I know. And I know that your play was peculiar. You were scamming me. And I'll figure out how".

"Well, be that as it may, I haven't the time to hang around arguing with you, John. And I am certainly not going to try to change your mind about anything. You go ahead and believe what you want to believe, and if you conclude that you managed to beat a scam, power to you. I have places to be and things to do, and must be heading off".

"Well, dammit, be that way. First you scam me then you brush me off. Some friend you are … Ohhh, wait, dammit, before I forget, same time next week? Or will you be busy scamming someone else at the time? Let me know such that I arrange my schedule properly".

"Sure, that works for me".

And Charlie the chess mate sped off to the next appointment, joining the traffic flow, absconding abruptly behind a row of shops.

References

Bainbridge, A. 2018, Autonomous vehicles & auto-tours: what is an auto-tour and how will autonomous vehicles impact tours, attractions & cities?, Autoura, The Spontaneous Travel Company Ltd, Courtenay, Oxfordshire, UK.

Beyer, A. 2016, Le transport fait-il partie du voyage? Pour une compréhension du déplacement touristique à partir de l'antagonisme contrainte/agrément, *Géotransports*, Vol. 7, pp. 7–22.

Billington, J. 2018, Kia to unveil future autonomous car that can read a passenger's emotions, *ADAS & Autonomous Vehicle International* magazine, December, https://www.autonomousvehicleinternational.com/news/adas/kia-to-unveil-future-autonomous-car-that-can-read-a-passengers-emotions.html.

Brook-Jones, C. 2022, BMW and AirConsole to integrate casual gaming system into 2023 vehicles, *ADAS & Autonomous Vehicle International* magazine, October, https://www.autonomousvehicleinternational.com/news/connectivity/bmw-and-airconsole-to-integrate-casual-gaming-system-into-2023-vehicles.html.

Cadillac Corporation 2023, SOCIALSPACE, Cadillac Corporation, https://www.cadillac.com/concept-vehicles/socialspace-concept.

Cohen, S.A. and Hopkins, D. 2019, Autonomous vehicles and the future of urban tourism, *Annals of Tourism Research*, Vol. 74, pp. 33–42.

Fujita, M. 2001, AIBO: toward the era of digital creatures, *The International Journal of Robotics Research*, Vol. 20, No. 10, pp. 781–794.

Hawkins, A.J. 2017, Toyota's Concept-i has built-in artificial intelligence named 'Yui', *The Verge* magazine, January, https://www.theverge.com/2017/1/4/14169960/toyota-concept-i-artificial-intelligence-yui.

Lakier, M., Nacke, L.E., Igarashi, T. and Vogel, D. 2019, Cross-car, multiplayer games for semi-autonomous driving, Proceedings of the Annual Symposium on Computer–Human Interaction in Play, Barcelona, Spain, 22–25 October, pp. 467–480.

Luka Inc. 2023, The AI companion who cares: always here to listen and talk, always on your side, Replika, https://replika.com.

Morby, A. 2017, Toyota's Concept-i car uses artificial intelligence to anticipate its driver's needs, *Dezeen* magazine, January, https://www.dezeen.com/2017/01/06/toyota-concept-i-car-anticipates-drivers-needs-ces.

Murphy, M. and Templin, J. 2017, This app is trying to replicate you, *Quartz* magazine, https://qz.com/1698337/replika-this-app-is-trying-to-replicate-you.

Rojas, R. In the heart of Nashville, rolling parties rage at every stoplight, *New York Times*, 20/9/2021, https://www.nytimes.com/2021/09/19/us/nashville-party-vehicles.html.

Schukert M. and Müller S. 2006, Erlebnisorientierung im touristischen transport am beispiel des personenluftverkehrs, in Weiermair K. and Brunner-Sperdin A., *Erlebnisinszenierung Im Tourismus*, Erich Schmidt Verlag, Berlin, Germany, pp. 153–166.

Shedroff N. and Noessel, C. 2012, *Make it so: interaction design lessons from science fiction*, Rosenfeld Media, Brooklyn, New York, New York, USA.

Takahashi, D. 2023, Nvidia lets you play cloud games in the car with GeForce Now expansion, *VentureBeat* magazine, January, https://venturebeat.com/games/nvidia-lets-you-play-cloud-games-in-the-car-with-geforce-now-expansion.

Woebot Health 2023, Small chats for big feelings, *Woebot*, https://woebothealth.com.

Chapter 10

Final Considerations

The previous chapters of this book have covered much ground. Many relevant facts have been cited and many references to sources of information have been made. The history of autonomous road vehicles was reviewed as were human-facing facts which will shape their design going forward. Autonomous road vehicles from the world of science fiction were also reviewed, highlighting what has already been said, and what has not yet been said, about them.

Five of the most frequently fielded future-facing speculative design approaches were discussed. And from among them, the science fiction prototype approach was selected as the basis for speculating about future autonomous road vehicles in this book. Sixteen science fiction prototypes of friendly neighbourhood robots were produced and presented which either transport people, provide a workplace, provide healthcare or provide entertainment.

Between them, the sixteen road vehicles exhibit many of the logical and ethical conundrums. which future designers will be facing. Many, possibly most, of the human-facing conceptual issues which will be encountered during the design of the future friendly neighbourhood robots have popped up somewhere among the previous pages of this book.

Having provided many relevant facts this final chapter now turns to a few final considerations of a strategic nature. These final points are a reflection of the topics, themes and techniques which have appeared previously in the book. But, unlike most of the material of the earlier chapters, not all of the observations made in this chapter are obvious or inevitable.

From the previous chapters the implications of autonomy on physical matters such as driving, parking, accepting inputs, providing outputs and dealing with malfunctions or breakdowns should now be somewhat obvious. And while possibly not as obvious, the implications of autonomy on the more psychological and sociological matters such as

DOI: 10.4324/9781032724232-10

the naturalness of the interaction, perceived agency and perceived trust should also be discernible.

The observations which follow in this chapter are instead largely in relation to issues which were not covered explicitly in the preceding chapters, and which are somewhat less obvious and somewhat further removed from the machine itself. They are the result of stepping back, viewing from a greater distance, and connecting a few of the hazy dots which can already be seen. The remaining material can be considered to be calls for care in relation to the design of the future friendly neighbourhood robots. And, in a few cases, they may even be warnings.

With Friendly Neighbourhood Robots What Changes?

The well-known science fiction author Douglas Adams (2002) famously wrote:

> I've come up with a set of rules that describe our reactions to technologies:
>
> 1   Anything that is in the world when you're born is normal and ordinary and is just a natural part of the way the world works.
> 2   Anything that's invented between when you're fifteen and thirty-five is new and exciting and revolutionary and you can probably get a career in it.
> 3   Anything invented after you're thirty-five is against the natural order of things.

If Adams is indeed correct, then for many of the people in the world today the friendly neighbourhood robots will prove to be very much "against the natural order of things". Unfamiliar, unpredictable and unnatural. The obvious opportunities and hopeful horizons of the future friendly neighbourhood robots have little in common with the incremental progress of human-driven road vehicles experienced by people during the 20th century. From current frames of reference, and adopting current societal norms, the friendly neighbourhood robots will be "against the natural order of things".

So far in this book many facts have been cited in relation to how friendly neighbourhood robots will drive themselves, provide services, talk to people and even respond to them emotionally. Some of what will be "against the natural order of things" in such areas has possibly already emerged. But there are also other aspects which are somewhat less immediate and somewhat less obvious.

One characteristic which will be "against the natural order of things", at least initially, is the cost. This is a point which is often neglected in discussions about future autonomous road vehicles. Logic suggests that they will have to provide most or all of the capabilities of current human-driven vehicles, but also many more. And changes in capability inevitably bring with them changes in cost. So many additional sensors and so much additional computer processing capability will invariably come at a price, literally. Given the development, production and maintenance costs of such complex machines the friendly neighbourhood robots will almost certainly be expensive robots.

In fact, much, possibly most, of the real-world value associated with a traditional human-driven road vehicle is not derived from the capabilities of the vehicle but instead from those of the driver. What is the value of a cab without a cabbie? What is the value of a police car without a police officer? And what is the value of a hatchback to a family whose members do not possess driving licences? Not all that much. And the rather reasonable cost of many current human-driven road vehicles is the confirmation. In part, the cost of 20th-century human-driven road vehicles was contained by the manufacturers by continuously shifting the increasingly complex driving decisions and other activities on to the human driver.

Instead, the friendly neighbourhood robots will have to do all of these things by themselves. Autonomous, independent and responsible. Such disruptively different forms of road transport will carry a premium price tag, will require a highly efficient business model and will require careful customer-centric design to get the sums to add up. And, above all, as Murphy (2019) has noted, "...a robot that violates the expectations of human ethics is subject to both rejection by consumers and product liability lawsuits".

Another characteristic of the friendly neighbourhood robots which will be "against the natural order of things", at least initially, will be its value to people. The term "value" can refer to the amount of money something costs, how useful it is to someone or its importance to someone. The greater cost has already been noted, and differences in usefulness and importance are also likely. In all three areas the friendly neighbourhood robots will be something different from traditional human-driven road vehicles.

Many different approaches to evaluating the usefulness of an artefact can be found in the fields of engineering, economics, philosophy, sociology and others. But regardless of which of the approaches is chosen when comparing the usefulness of a human-driven road vehicle to a friendly neighbourhood robot, it is almost certain that the results will be different.

Even when manifested in the form of a robo-taxi for short trips from point A to point B, the friendly neighbourhood robot will be providing much information and many services along the way. People will leverage the robot's information handling capabilities and automation in relation to many decision-making matters. This single difference is already sufficient to suggest a much wider field of utility than that of human-driven road vehicles.

And it should be noted that the friendly neighbourhood robots will usually act as a service rather than act as a possession. In most cases the value will be provided by something which the friendly neighbourhood robot does, rather than something which it is. Holbrook (1999) has suggested eight types of consumer value: efficiency, excellence, status, esteem, play, aesthetics, ethics and spirituality. And a quick thought experiment suggests that whereas a personally owned automobile can provide some value in most or all of Holbrook's categories, a mobility service will end up providing value in only a subset of the categories such as efficiency, excellence, play and ethics.

Another characteristic of the friendly neighbourhood robots which will be "against the natural order of things", at least initially, is their legal liability. While not completely different from the situation with human-driven road vehicles, friendly neighbourhood robots pose several new and distinct challenges. Autonomy brings with it both responsibility and liability.

The poor quality of many roads, the inefficiencies of urban traffic management systems, the complexities of human-driven road vehicles and the dangers of speed have all conspired over the years to produce a vast number of road accidents and road deaths. The large variety of accidents and their associated forms of harm and injury have posed challenges to insurance and legal systems since the arrival of substantial numbers of motor vehicles in the early 20th century.

Despite the difficulties, however, logical simplifications have been deployed to render the task of assigning culpability and determining liability more systematic. For example, the distinction between a mechanical failure of the vehicle or a driving error on the part of the human has been a constant from the time of introduction of the first automotive insurance policies. Separating the legal responsibilities of the vehicle manufacturer from those of the vehicle owner has been a necessity. And despite the occasional borderline case, the separation has not been too difficult to achieve in practice.

But the introduction of the friendly neighbourhood robots complicates matters with respect to the traditional logic and to the historical precedent. The friendly neighbourhood robots introduce new characteristics and new capabilities in many areas. Their manner of interacting with humans, their degree of compliance with socially accepted norms of the road and their need to defend themselves from misuse or harm all raise new legal and ethical questions. And require new legal definitions and codes.

For example, who exactly is responsible for causing the accident when a friendly neighbourhood robot opens its door into the path of a runner on the sidewalk, causing a collision which injures the runner? Or when the sun's glare on a bright afternoon causes a pedestrian to not see the robot's turn signal indication, and there is no human driver who can provide the backup redundancy? Or when a robot fails to detect a major illness of its only passenger, thus failing to call for medical assistance?

As authoritatively described elsewhere (see, for example, Channon et al. 2019 and Turner 2018) many changes will have to be made to existing legislation such as The Automated and Electric Vehicles Act 2018 (Marson et al. 2020) in order to provide appropriate legal frameworks for the insurance status and product liability of the friendly neighbourhood robots. Clarifications are needed in several areas which were not previously relevant with human-driven road vehicles. Such as the robot's situational awareness, signalling behaviours, passenger supports and self-defence mechanisms.

New legislations and new insurance industry practices will have to resolve a number of thorny issues in logic in relation to determining culpability and liability. For example, a precise definition will be needed of the concept of artificial intelligence and of the exact role which it performs in driving the vehicle. Precise definitions will also be needed of exactly what constitutes road data, what constitutes vehicle data and what constitutes the data provided explicitly or implicitly by the human passengers. And the new legislations and new insurance industry practices will have to define the exact role which each form of data has in making the driving decisions and ensuring the driving safety. For example, when a first-time tourist requests to be taken to a location which has just been subject to major flooding and some incident occurs, is the liability for the damages falling to the vehicle manufacturer, the mapping service provider, the taxi service provider or the tourist who requested the location?

And beyond the driving safety, new legislations and new insurance industry practices will have to be devised to deal with the many non-driving services which friendly neighbourhood robots will be providing. Robots which provide workplace capabilities may need insurance coverage which is analogous to that required for similar brick-and-mortar offices or factories. And robots which provide medical or entertainment services will presumably need to be handled in a somewhat similar manner to analogous fixed location installations. There appears to be much work to do to clarify how a service increases the risks beyond those associated with the traditional transport function. Or, stated the other way around, how the mobility adds to the risks normally associated with the service.

And, possibly most importantly, legal and insurance systems will have to be made compatible with any legislations which are enacted in relation to robot rights (see Gunkel 2018 for a review). There is growing recognition that some legal rights will be assigned to friendly neighbourhood robots as had happened with humans, then animals, and most recently with some elements of the natural environment. Whether due to the machines reaching a degree of consciousness (Will Theory) or due instead to the effects on human society of how we treat them (Interest Theory), some moral and some legal recognition of some robot rights seems imminent. Such legal rights will provide a cornerstone for determining what the friendly neighbourhood robots can do, what they can't do and who might be responsible when something goes wrong.

What Challenges Will the Designers of the Friendly Neighbourhood Robots Face?

The future friendly neighbourhood robots will exhibit many new characteristics and capabilities. They will challenge our understanding of what constitutes a machine, what instead constitutes a form of life and what we as humans desire from the world. As Krippendorff (2006) has suggested, "humans do not see and act on the physical qualities of things, but on what they mean to them". Thus the design of the future friendly neighbourhood robots will be more about how we humans wish to live than about what the technology can do.

Explicitly or implicitly the pages of this book have suggested many technical, social and ethical challenges. The known facts in relation to aesthetics, dynamics, behaviours, conversations, personalities and trust suggest that many design choices will reveal themselves to have major implications for people due to the degree of autonomy bestowed upon the vehicles. Many tricky decisions are lying in wait for the designers of the future friendly neighbourhood robots.

The science fiction prototypes of Chapters 6 through 9 in particular have suggested numerous challenges and nuances. From difficulties of categorisation to the degree of companionship, they have raised questions which will soon need to be answered. The situations which are tested today as science fiction prototypes will in the future need to be tested as real world prototypes. They are a small sample of what is to come.

Several themes can be noted across the sixteen science fiction prototypes, with some appearing in more than a single story. Table 10.1 lists the most obvious themes which the reader will probably have spotted. On the left are the themes and on the right are the science fiction prototypes which most directly and most obviously highlighted them.

TABLE 10.1 Themes explored by the science fiction prototypes of Chapters 6 through 9.

carrying capacity:	Marrakesh Making, Plumber Jim, Roman
on-board equipment:	HFT1, Marrakesh Making, Medical Transfer Shuttle 42, Plumber Jim
dealing with roads and weather:	Roman, Tovar Tester
dealing with gaps in network coverage:	Ace Taxis, HFT1
dealing with hacking or misuse:	Fans R Us, Privacy Pod
dealing with human requests:	Ace Taxis, Dustoff Danny, Fans R Us, Roman
dealing with emergencies:	Ace Taxis, Dustoff Danny, Medical Transfer Shuttle 42
human backup:	Ace Taxis, Allride, Roman
anthropomorphism and autonomy:	Dave's Diary, Dustoff Danny
linguistic abilities:	Big Easy Funerals, Chess Mate, Happy Valley Health, Roman, Social Sara
emotional abilities:	Big Easy Funerals, Chess Mate, Happy Valley Health, Roman, Social Sara
cultural awareness:	Big Easy Funerals, Chess Mate, Happy Valley Health, Roman, Social Sara
companionship and friendship:	Chess Mate
workplace provision:	Marrakesh Making, Privacy Pod
healthcare provision:	Medical Transfer Shuttle 42, Tovar Tester
social care provision:	Chess Mate, Happy Valley Health
entertainment provision	Chess Mate, Roman, Privacy Pod
certification:	Happy Valley Health, Medical Transfer Shuttle 42
safety:	Ace Taxis, AllRide, Dustoff Danny, Fans R Us, Medical Transfer Shuttle 42
ethics:	AllRide, Chess Mate, Privacy Pod

As can be noted from Table 10.1, the science fiction prototypes touch upon themes ranging from the physical to the social and normative. While each merits detailed consideration, the discussion here will be limited to those which appear to be the most problematic from today's perspective.

As highlighted by "Happy Valley Health" and "Medical Transfer Shuttle 42" there is an important issue of categorisation and certification. The capable computerised companions will perform the multiple tasks of a single human, and perhaps the multiple tasks of multiple humans. Such a wide variety of capabilities and services leads naturally to difficulties in classification and certification. In a world where the automated systems will be capable of performing many different human activities, how will the robots be divided up so as categorise their function and provide the basis for formal certification? Just as humans train to perform certain roles in society and must pass certain tests before they are allowed to do so, the friendly neighbourhood robots will also need to be classified and certified.

In both the "Happy Valley Health" and the "Medical Transfer Shuttle 42" stories the friendly neighbourhood robot is delegated a set of significant responsibilities which would traditionally have fallen to one or more humans. In both cases the exact realm of the physical and psychological support which was on offer could have been better clarified. And in both cases the functions and authority involved a mixture of what would traditionally have been assigned to the road vehicle and what would traditionally have been assigned to the one or more humans.

The world of traditional human-driven road vehicles provides some examples of systems of classification such as the international standards in relation to ambulances, emergency vehicles, firefighting vehicles and other specialised items of equipment. However, such traditional systems of classification usually do not extend beyond the vehicle itself to also include the human operators. Existing systems of classification do not usually cover the full range of capabilities and responsibilities which the future friendly neighbourhood robots will be providing.

In the cases of "Happy Valley Health" and "Medical Transfer Shuttle 42" any shortcomings or operating errors on the part of the robot can lead to dramatic consequences for the users of the service. The possible risks to mental health can be imagined in one case, and an actual death occurred in the other. Thus detailed systems of classification seem needed and some degree of certification seems inevitable. Autonomy brings with it both responsibility and liability.

As highlighted instead by "Dave's Diary" and "HFT1" there are also questions to be answered in relation to the degree of autonomy to assign to a friendly neighbourhood robot, and in relation to how that autonomy is to be communicated to people. Designers will have to clearly identify who is responsible for what and who is in charge of what. The structure of the team formed by the friendly neighbourhood robot and the humans will need to be made fully obvious to all involved.

In "Dave's Diary" psychological and sociological pressures ensue from a work relationship with the automation which is not fully and obviously defined. The automation is sufficient to render the human work tedious and unsatisfying most of the time, but exits from its responsibilities when something unexpected or unusual occurs. The human is thus required to sporadically transition from situations of little or no responsibility to situations of near complete responsibility, with all the associated stresses and pressures. "Dave's Diary" highlights how human–machine relationships have traditionally been organised along the functional lines of what each of the partners can do, rather than along the psychological and motivational lines of what each of the partners expects or is comfortable assuming responsibility for.

The "HFT1" story instead sees the automation taking on representational and decision-making roles which require boundaries, settings and uninterrupted real-time coordination to perform properly. In "HFT1" the automation's reliability and usefulness depended on control parameters which were not always possible to fully guarantee, raising questions about where the capabilities can be safely deployed and whether the capabilities should be allowed by law in the first place. The "HFT1" story has a friendly neighbourhood robot replacing its user for certain business transactions due to the power of its on-board computational hardware, but while technically possible, and ultimately successful in the story, such representational delegation comes with obvious risks of a financial and ethical nature.

"Big Easy Funerals", "Happy Valley Health", "Roman" and "Social Sara" highlighted instead issues of cultural awareness. As noted in the stories, other road users, passengers and pedestrians can become annoyed or even upset by interactions which are not informed by linguistic and cultural common ground.

If a friendly neighbourhood robot is intended to only transport people or goods for short distances under simplified conditions, then perhaps linguistic and cultural awareness is not a needed characteristic. For example, driving on a busy motorway involves several elements of social interaction with other human drivers, but those simple interactions can be mimicked by the friendly neighbourhood robot in a somewhat mechanical manner.

But if the robot is of a type which must also provide some form of service which involves the linguistic, social or cultural, then a degree of linguistic and cultural awareness becomes important to ensure proper operation. And if the journey time extends to lengthy periods of hours or days of interaction and co-dependence, then the robot's linguistic and cultural awareness in relation to its human passengers may prove decisive. In such cases the common ground with humans will need to be built into the friendly neighbourhood robot either explicitly as in the case of formal rules in code or implicitly as in the case of the selection of the datasets used for the machine learning.

As "Big Easy Funerals", "Happy Valley Health", "Roman" and "Social Sara" suggest there may be many future friendly neighbourhood robots which perform duties which involve a significant number of linguistic interactions of a ritualistic or social nature. With an obvious need for a degree of cultural awareness and common ground. Not identifying people correctly and not understanding what is appropriate to say to them proves to be a major issue in the "Big Easy Funerals" story. Asking inappropriate questions and stimulating unpleasant memories is an obvious risk in both the "Happy Valley Health" and "Social Sara" stories. And an inability to fully understand passenger requests or to communicate culturally sensitive policies is an obvious impediment to the passenger experience in "Roman".

Finally, as highlighted by "Chess Mate", there is already today a growing societal concern in relation to the degree of companionship or even friendship which robots provide. Either intentionally or unintentionally. If a friendly neighbourhood robot is intended to only transport people or goods for short distances under simplified conditions, then perhaps there is little risk of bonds forming and of the robot being seen as a companion or a friend. But if the robot is of a type which must also provide some form of service which involves the linguistic, social or cultural, then there will be a substantial risk of bonds forming and of impressions of companionship or even friendship arising.

On the one hand friendly neighbourhood robots might benefit people by alleviating feelings of boredom, reducing loneliness and providing emotional support. For some people, interacting with a robot may prove to be a source of comfort and wellbeing. In public discourse there have been many functions and roles suggested for the future friendly neighbourhood robots where a degree of linguistic, social or cultural interaction would improve the customer experience.

But on the other hand there are also concerns that relying on friendly neighbourhood robots could potentially hinder or replace human-to-human interactions. If people become overly dependent on them it could lead to decreased social interaction with other humans, exacerbating feelings of isolation and making it more difficult to establish meaningful human relationships. Over the years several terms have been coined to describe the potential risks including automation dependency, machine addiction, machine reliance, robotic camaraderie, robot reliance and techno-dependency. The existence of so many different terms to describe the risk suggests the many facets of the matter and the degree of public concern.

And it should be noted that friendship with robots may indeed be possible. Danaher (2019), for example, has argued that science fiction has explored the concept of friendship with alien beings which were dramatically different from humans. And that the public reactions to the stories support the view that such friendships are a real possibility. And if readers can empathise, sympathise and identify with science fiction creatures which have little or no biological, linguistic or cultural common ground with humans, then why would they not react favourably to friendly neighbourhood robots which were designed specifically for them?

Evidence from fields including neuroscience, psychology and sociology suggests that there is currently little support for the view that friendship is speciesist. And if the coming years should confirm that there are few logical or philosophical impediments to friendships with robots, then it will be the designers who will have to suggest the degree of friendship (utility, pleasure or virtue) which is consistent with the specific friendly neighbourhood robot. Targets for such matters may be needed early in the concept definition phase to constrain the design process and to provide a reference for the nature and degree of anthropomorphism to adopt for the vehicle.

When Might the Friendly Neighbourhood Robots Arrive on Our Streets?

When might the friendly neighbourhood robots arrive on our streets? This is a question which is on the mind of many. It is a deceptively simple question, but is not one which is straightforward to answer. And may not even be the right question to ask.

The reason why it may not be the right question to ask has to do with the implicit assumptions involved. With the definition of "friendly neighbourhood robot" in mind (see Chapter 1) it can be stated without reservation that there will be many different machines which will come under that general heading. There is currently no standard specification for a friendly neighbourhood robot, and such a standard is unlikely to emerge in the near future given the variety of technologies and roles which such sophisticated machines might involve. Speaking of the arrival of the friendly neighbourhood robots may therefore make little sense since different machines with different capabilities will arrive at different times.

Inevitably, any designed artefact enters the world based on the simplest available technology which meets the requirement. It then becomes progressively more complex and more sophisticated as new technological opportunities and new requirements arise. Humble beginnings always give way to far less humble destinies. It would thus seem reasonable to consider the arrival of the friendly neighbourhood robots to be a process involving successive waves of machines, starting from the simplest and least technically challenging then progressing onwards until eventually arriving at lifelike complexity.

Figure 10.1 provides one response to the question of when the friendly neighbourhood robots will arrive on our roads. The timeline is based on logical deduction from the known facts about the technologies involved and from discussions with industry experts. And the timeline is consistent with the idea of successive waves of increasingly more complex machines.

The timeline assumes that the first friendly neighbourhood robots will be fully self-driving but may not offer much more in terms of their automation. The capabilities of existing robo-taxis will be widened by new visual, linguistic and communicative abilities, but little more. The early robots will have some interactions and some behaviours which are complex and which may thus appear to exhibit a degree of understanding and intention, but none will exist. Like philosophical zombies (Kirk 2019), their behaviour will be the result of fixed programming and simulation rather than informed acts of a semi-conscious or conscious nature.

The first friendly neighbourhood robots will provide basic point A to point B transport of people or goods and are likely to start becoming

Full Self-Driving on Demand   **Business, Healthcare and Entertainment Services**   **Full Conversational and Emotional Capabilities**

Time

Figure 10.1 Proposed timeline of introduction of the friendly neighbourhood robots.

common in cities and on motorways during the period from 2025 to 2030. Given the current acceleration in self-driving technology it is difficult to imagine the first commercially viable robots arriving much later than 2030. Though different from traditional human-driven road vehicles, the first robots will nevertheless have many components and many characteristics which are obvious products of the automotive tradition.

After the arrival of fully self-driving road vehicles which can operate for extended periods of time without human intervention, the next friendly neighbourhood robots which are likely to appear will provide business, healthcare or entertainment services. At some point between 2030 and 2035 we should begin to see the first friendly neighbourhood robots which are specifically designed as business, healthcare or entertainment machines. Providing specific aspects of the existing bricks-and-mortar based services in a more flexible and mobile manner. And at that point the friendly neighbourhood robots will start to accelerate their evolution away from being a form of transport to being instead a form of service.

The services will not be grafted on to, but instead designed into, these autonomous road vehicles. The robots will involve not simply speciality software but also specialised interfaces, speciality cabin layouts and unique vehicle packaging arrangements. They will blend the transportation capabilities of road vehicles with some of the specialist functions and technologies of some services which have traditionally been provided by humans. The many current proposals for mobile offices, healthcare clinics and entertainment centres suggest that such robots will arrive on the scene once full self-driving is safe and reliable. With the self-driving fully achieved, the automation will be applied more widely to provide further services which are of benefit to people.

And, finally, as also suggested in Figure 10.1 there is the inevitable end point of lifelike friendly neighbourhood robots which are fluent in language, adept at conversation, emotionally aware and emotionally responsive. By about 2035 or 2040 some friendly neighbourhood robots will appear on the scene which to many people will seem to be a new form of life rather than a new form of transport. Such friendly neighbourhood robots will be designed to have electronic systems which are more biologically inspired than the current generation of engineering systems, and will raise a multitude of philosophical and ethical questions in relation to their design and deployment.

Such robots will interact with humans based on a deep understanding of human biology, language, emotion and common ground. They will provide specialised services of a technically, psychologically and sociologically complex nature. Design and engineering support may be provided by some, while filling in as real-time manufacturing facilities will be the role of others. Some of the robots will replace brick-and-mortar buildings as meeting places, service centres or shops, while others will act as lifestyle assistants, entertainment centres or holiday cruisers. There will be gofers, valets and companions. And there will be robots which perform medical tests, offer emergency transport or even replace junior doctors in the case of minor mishaps. The fully conversational and fully emotional robots will perform many repetitive tasks and take on many minor responsibilities in support of people. The societal challenge, if one exists, is simply that of choosing which.

Will the Friendly Neighbourhood Robots Be Considered a New Form of Life?

The question of whether the friendly neighbourhood robots will be considered a new form of life is not straightforward to answer because proponents for or against the proposition tend to adopt different definitions of life. Nevertheless, despite the difficulties, the question is worth pursuing because it has important design implications. How a person reacts to a friendly neighbourhood robot being late in arriving, or making a mistake during its service provision, will depend greatly on whether the person considers it to be a simple machine which executes a rigid function or instead a complex machine which is capable of making decisions. Intent matters.

One of the earliest attempts at defining life was that of Aristotle. In *De Anima* of 350 BCE (Shields 2016) he listed four properties which he considered essential for life:

- reproduction;
- growth;
- sensation;
- self-movement.

Based on Aristotle's account any form of robot, including the friendly neighbourhood kind, would appear to fall foul of the reproduction requirement since robots often have some capacity for self-repair but usually do not have the ability to self-replicate. Nevertheless, there is some consensus that there are no theoretical limits to machine self-replication (see, for example, Von Neumann 1966 and Wikipedia Contributors 2023). Indeed, there have been several proposals over the years for specialist self-replicating robots (see, for example, Chirikjian et al. 2002; Zykov et al. 2005; Ellery 2016). It is therefore not beyond the realm of possibility that full self-replication may become a feature of some friendly neighbourhood robots at some point in the future. It would thus seem premature to exclude friendly neighbourhood robots from the realm of life forms based only on Aristotle's properties.

More recent times have seen a proliferation of attempts at defining life due to the increasing need to deal with viruses and other borderline cases in the biological realm, and the growing number of computer algorithms and robotic systems in the artificial intelligence realm. One influential example is that of the seven pillars, or PICERAS principles, proposed by Koshland (2002). They are suggested to be necessary conditions required for any living system to scientifically satisfy its kinetic and thermodynamic constraints:

- A programme: an organized plan that describes both the ingredients themselves and the kinetics of the interactions among ingredients as the living system persists through time.
- Improvisation: because a living system will inevitably be a small fraction of the larger universe in which it lives, it will not be able to control all the changes and vicissitudes of its environment, so it must have some way to change its program.
- Compartmentalisation: all the organisms that we consider living are confined to a limited volume, surrounded by a surface that we call a membrane or skin that keeps the ingredients in a defined volume and keeps deleterious chemicals, toxic or diluting, on the outside.
- Energy: life as we know it involves movement of chemicals, and a system with net movement cannot be in equilibrium. It must be an open and, in this case, metabolising system.
- Regeneration: because a metabolising system composed of cata-lysts and chemicals in a container is constantly reacting, it will inevitably be associated with some thermodynamic losses. Because those losses will eventually change the kinetics of the program adversely, there must be a plan to compensate for those losses, that is, a regeneration system.
- Adaptability: bodies respond to depletion of nutrients (energy supplies) with hunger which causes food seeking. And bodies prevent from eating excess nutrients by losing appetite. Walking long distances on bare feet leads to calluses to protect the feet. These behavioural manifestations of adaptability are a development of feedback and feedforward responses at the molecular level and are responses of living systems that allow survival in quickly changing environments.
- Seclusion: in a metabolising system with many reactions occurring at the same time it is essential to prevent the chemicals in pathway

1 from being metabolised by the catalysts of pathway 2. Living systems do this by the specificity of enzymes which work only on the molecules for which they were designed and are not confused by collisions with miscellaneous molecules from other pathways.

Application of Koshland's pillars to autonomous robots would seem to suggest that they meet the criteria. Relatively straightforward arguments can be made in support of the robots for each of the seven principles since they normally have components and programmes which compartmentalise, seclude, regenerate, utilise energy and adapt. Based on Koshland's criteria, friendly neighbourhood robots may be a new form of life.

Along somewhat similar lines, Cheok and Zhang (2019) compared recent robots to the criteria established by American biologist James Grier Miller in his influential 1978 book *Living Systems*. Miller defined his set of critical subsystems based on function rather than on their physical characteristics or biological attributes, thus achieving a description (see Table 10.2) which was independent of the material substrate or construction details.

TABLE 10.2 Miller's twenty critical subsystems which define life.

The Reproducer	The subsystem which is capable of giving rise to other systems similar to the one it is in.
The Boundary	The subsystem at the perimeter of a system that holds together the components which make up the system, protects them from environmental stresses, and excludes or permits entry to various sorts of matter-energy and information.
The Ingestor	The subsystem which brings matter-energy across the system boundary from the environment.
The Distributor	The subsystem which carries inputs from outside the system, or outputs from its subsystem around the system to each component.
The Converter	The subsystem which changes certain inputs to the system into forms more useful for the special processes of that particular system.
The Producer	The subsystem which forms stable associations that endure for significant periods among matter-energy inputs to the system or outputs from its converter, the materials synthesised being for growth, damage repair or replacement of components of the system, or for providing

(Continued)

TABLE 10.2 Cont.

	energy for moving or constituting the system's output of products or information markers to its suprasystem.
Matter-energy storage	The subsystem which places matter or energy at some location in the system, retains it over time and retrieves it.
The Extruder	The subsystem which transmits matter-energy out of the system in the forms of products or waste.
The Motor	The subsystem which moves the system or parts of it in relation to part or all of its environment or moves components of its environment in relation to each other.
The Supporter	The subsystem which maintains the proper spatial relationships among components of the system, so that they can interact without weighing each other down or crowding each other.
The Input Transducer	The sensory subsystem which brings markers bearing information into the system, changing them to other matter-energy forms suitable for transmission within it.
The Internal Transducer	The sensory subsystem which receives, from subsystems or components within the system, markers bearing information about significant alterations in those subsystems or components, changing them to other matter-energy forms of a sort which can be transmitted within it.
Channel and Net	The subsystem composed of a single route in physical space, or multiple interconnecting routes, over which markers bearing information are transmitted to all parts of the system.
The Timer	The subsystem which transmits to the decider information about time-related states of the environment or of components of the system. This information signals the decider of the system or deciders of subsystems to start, stop, alter the rate, or advance or delay the phase of one or more of the system's processes, thus coordinating them in time.
The Decoder	The subsystem which alters the code of information input to it, through the input transducer or internal transducer, into a "private" code that can be used internally by the system.
The Associator	The Associator is the subsystem which carries out the first stage of the learning process, forming enduring associations among items of information in the system.

(Continued)

TABLE 10.2 Cont.

The Memory	The Memory is the subsystem which carries out the second stage of the learning process, storing information in the system for different periods of time, and then retrieving it.
The Decider	The executive subsystem which receives information inputs from all other subsystems and transmits to them information outputs for guidance, coordination and control of the system.
The Encoder	The subsystem which alters the code of information input to it from other information-processing subsystems, from a "private" code used internally by the system into a "public" code which can be interpreted by other systems in its environment.
The Output Transducer	The subsystem which puts out markers bearing information from the systems, changing markers within the system into other matter energy forms which can be transmitted over channels in the system's environment.

Cheok and Zhang's comparison proved positive. In their words:

> I claim that the brief comparisons I make between Miller's 20 subsystems and various of the characteristics of robots, provide sufficient justification for us to consider robots to be living systems according to Miller's 1978 theory. So if we accept Miller's theory, and to the best of my knowledge it has never been refuted, we can safely answer our title question in the affirmative. Yes, robots are alive.

And an approach based on an even greater degree of abstraction away from everyday examples of biological life was suggested by Çengel (2022). Starting from known physical, chemical and thermodynamic laws of nature, he proposed that all forms of life respect six core propositions:

- Life exists. It just does, just like the law and force of gravity exist. Life belongs to the same category of existence as the causal laws of physics.
- Life is a field phenomenon, like quantum fields, except that the domain of influence of the life field is the animate realm and not the entire spacetime.

- Life is an anomaly in the physical realm since its existence cannot be predicted by the laws and forces of physics and the cause–effect relations.
- What is an inanimate being today can be part of the body of a living being tomorrow and become animate.
- Life is an agency with causal power, rather than an emergent property of assemblies of matter that just characterises the assembly. Life qualifies as an active agent that rules matter and fully governs the material content within its sphere of influence.
- The capacity to die is a unique, distinctive and characteristic feature of living beings. A thing that cannot experience death cannot be alive.

Çengel summarised his position as "A natural entity whose internal changes and external behavior cannot be predicted by the universal laws and forces of physics alone at all times is an animate or living being. Everything else is an inanimate or nonliving being". And from this position and point of view arrived at a negative conclusion in relation to robots. In his own words:

> From battery-powered toys to robots, smart phones and autonomous vehicles, all manmade devices that are powered by externally-supplied energy such as fuels and electricity are lifeless since the operation of those devices comes to a complete halt when they run out of the externally supplied energy. The behavior of a robot with a completely discharged battery, for example, can be fully predicted by the universal laws and forces of physics. When the battery is recharged and the robot is turned on, the robot will act in accordance with the laws and influences encoded in its software, which constitute the artificial agency of robot. The behavior of living being is also determined solely by the laws and forces of physics when they run out of energy and die. But unlike smart devices that 'die' metaphorically when they run out of energy, a living being can never be brought back to life once it dies, no matter how much energy we supply. As such, the natural agency of life is characteristically different from the artificial agency of robots and other smart devices.

Independent of the philosophical and scientific considerations, however, there remains the everyday reality of people's subjective experiences of interacting with friendly neighbourhood robots. Whether officially classified as a life form by authorities or not, people may judge them to be life forms based on impressions from interactions and based on the roles which the robots fulfil in society. And, as discussed elsewhere (see Giacomin 2023), humans have a great propensity for anthropomorphising natural phenomena, animals and machines.

Anthropomorphism refers to the attribution of human traits, emotions or intentions to non-human entities with a view to rationalising their actions. It is an innate human tendency which becomes increasingly more stimulated with increasing levels of behavioural complexity. Humans have anthropomorphised their interactions with animals and their descriptions of them from early times. Evidence also exists for the anthropomorphising of machines, or at least for the anthropomorphising of those machines which were sufficiently multifunctional and complex to appear to human eyes to be exhibiting behaviours. And much evidence suggests that people's interactions with road vehicles and their opinions about them tend to be influenced by the available anthropomorphic cues.

Friendly neighbourhood robots which are fluent in language, adept at conversation, emotionally aware and emotionally responsive will almost certainly stimulate the human anthropomorphising tendency. Indeed, autonomous road vehicles have already been the subject of several studies (see Giacomin 2023 for an introduction) which have investigated the effects of anthropomorphic cues on subjective responses such as people's perceptions of the vehicle's agency, trust and friendship. And in all the cases the anthropomorphic cues were found to have a substantial effect, with the people's perceptions increasing with the increasing anthropomorphising of the vehicle.

Empirical evidence thus suggests that people's impressions from interactions with friendly neighbourhood robots, and from the roles which the robots fulfil, are highly anthropomorphic in nature. People appear better able to make sense of what a robot is doing if they think about the machine in anthropomorphic terms, i.e. as a form of life.

It can thus be argued that the existing empirical evidence suggests that many people are thinking about friendly neighbourhood robots more as a new form of life than as a new form of transport. A new form of life whose characteristics, capabilities, metaphors and meanings will have to be chosen with great care by the designers.

References

Adams, D. 2002, *The Salmon Of Doubt: hitchhiking the universe one last time*, Vol. 3, Macmillan Publishers, New York, New York, USA.

Çengel, Y.A. 2022, A novel theory of life and its implications on viruses and robots, *Journal of Future Robot Life*, Vol. 3, No. 2, pp. 183–205.

Çengel, Y.A. 2023, Eighteen distinctive characteristics of life, *Heliyon*, Vol. 9, No. 3, e13603.

Channon, M., McCormick, L. and Noussia, K. 2019, *The Law and Autonomous Vehicles*, Informa Law from Routledge, Abingdon, Oxon, UK.

Cheok, A.D. and Zhang, E.Y. 2019, Are robots alive?, In A.D. Cheok and E.Y. Zhang (eds) *Human–Robot Intimate Relationships, Human–Computer Interaction Series*, Springer Nature Switzerland AG, Cham, Switzerland, pp. 159–188.

Chirikjian, G.S., Zhou, Y. and Suthakorn, J. 2002, Self-replicating robots for lunar development, *IEEE/ASME Transactions on Mechatronics*, Vol. 7, No. 4, pp. 462–472.

Danaher, J. 2019, The philosophical case for robot friendship, *Journal of Posthuman Studies*, Vol. 3, No. 1, pp. 5–24.

Ellery, A. 2016, Are self-replicating machines feasible?, *Journal of Spacecraft and Rockets*, Vol. 53, No. 2, pp. 317–327.

Giacomin, J. 2023, *Humans and Autonomous Vehicles*, Routledge, London, UK.

Gunkel D.J. 2018, *Robot Rights*, MIT Press, Cambridge, Massachusetts, USA.

Holbrook, M.B. 1999, *Consumer Value: a framework for analysis and research*, Routledge, Abingdon, UK.

Kirk, R. 2019, Zombie, *The Stanford Encyclopedia of Philosophy*, E.N. Zalta (ed.), Summer Edition, https://plato.stanford.edu/entries/zombies.

Koshland, D.E. 2002, The seven pillars of life, *Science*, Vol. 295, No. 5563, pp. 2215–2216.

Krippendorff, K. 2006, *The Semantic Turn: a new foundation for design*, Taylor & Francis, Boca Raton. Florida, USA.

Marson, J., Ferris, K. and Dickinson, J. 2020, The Automated and Electric Vehicles Act 2018 Part 1 and Beyond: a critical review, *Statute Law Review*, Vol. 41, No. 3, pp. 395–416.

Miller J.G. 1978, *Living Systems*, McGraw-Hill, New York, New York, USA.

Murphy, R.R. 2019, Learn AI and human–robot interaction from Asimov's I, Robot stories, *Robotics through Science Fiction*, Vol. 2, Robin R. Murphy, Printed by Amazon, UK.

Shields, C.J. 2016, *Aristotle: De Anima*, Clarendon Aristotle Series, Oxford University Press, Oxford, Oxfordshire, UK.

Turner, J. 2018, *Robot Rules: regulating artificial intelligence*, Palgrave Macmillan, Cham, Switzerland.

Von Neumann, J. 1966, *Theory of Self-Reproducing Automata*, University of Illinois Press, Urbana, Illinois, USA.

Wikipedia Contributors 2023, Self-replicating machine, Wikipedia, The Free Encyclopedia, https://en.wikipedia.org/w/index.php?title=Self-replicating_machine&oldid=1138699200.

Zykov, V., Mytilinaios, E., Adams, B. and Lipson, H. 2005, Self-reproducing machines, *Nature*, Vol. 435, pp. 163–164.

Index

Note: Page numbers in **bold** indicate tables; those in *italics* indicate figures